DATE DUE		
NOV 1 9 2008			

AMONG THE
Cannibals

AMONG THE

Cannibals

ADVENTURES
ON THE TRAIL OF
MAN'S DARKEST RITUAL

Paul Raffaele

❁ Smithsonian Books

◐ **Collins**
An Imprint of HarperCollins*Publishers*

HarperCollins books may be purchased for educational, business, or sales promotional use. For information, please write: Special Markets Department, HarperCollins Publishers, 10 East 53rd Street, New York, NY 10022.

FIRST EDITION

Designed by Chris Welch

Printed on acid-free paper

Library of Congress Cataloging-in-Publication Data
Raffaele, Paul.
    Among the cannibals : adventures on the trail of man's darkest ritual / Paul Raffaele.—1st ed.
        p. cm.
    ISBN 978-0-06-135788-6
    1. Cannibalism. I. Title.
GN409.R35 2008
394'.9—dc22
2007050481

08 09 10 11 12   ID/RRD   10 9 8 7 6 5 4 3 2 1

TO CATHERINE AND CECILIA

CONTENTS

AMONG THE
Cannibals

PREFACE

Cannibals: one of the most compelling, and repelling, words in the English language. Of all humanity's known taboos through the millennia, eating human flesh is the most repugnant and forbidding. And, yet, the enormous worldwide popularity of the books and movies about Hannibal Lecter, the cannibal monster, shows how we are drawn irresistibly to tales about this deep and terrible perversity of human nature. Few of us ever want to be cannibals, but most of us want to hear stories about them.

Cannibalism comes in several forms. There are countless morbid true stories of psychopaths like Lecter, deeply aberrant people who defied social custom by indulging their hunger for human flesh. Recent examples include the Japanese Issei Sagawa, who ate his Parisian girlfriend in 1981, and Marc Sappington in Kansas, who axed to death a sixteen-year-old boy, and ate part of his leg in 2001. Jeffrey Dahmer of Milwaukee, from 1978 to 1991, killed and ate seventeen men. He claimed that he liked them so much that he ate them so they would become part of him.

Not all cannibals are psychopaths. In many warrior societies, fighters honored or insulted those they had slain by eating their flesh in ritual feasts, while in some rural hamlets of China and also in some Amazon villages people ate the body parts of beloved relatives to remember them. The most prevalent role of cannibalism throughout history has been as a powerful and pivotal part of religious ceremonies, but the only form that has ever proved socially acceptable to outsiders has been "lifeboat" cannibalism. This arises from the fierce will to survive when starving during famine or warfare, or when lost on land or at sea. One famed example followed the aftermath of the plane crash in the Andes of Uruguayan Air Force Flight 571 in 1972, when surviving passengers stayed alive until rescued by eating the flesh of passengers who died. This formed the plot for a major book, *Alive: The Story of the Andes Survivors*, by Piers Paul Read, published in 1974 and released as a motion picture in 1993.

Cannibalism has likely been with us for almost as long as humans have walked the Earth, even though most cultures have made the eating of human flesh taboo and condemned it as evil and demonic. "Worldwide folklore, oral traditions, sacred writings, anthropological narratives, war stories, urban police records, clinical psychological sources and tales of lost or helpless wanderers and explorers tell of cannibal peoples and cannibal events," wrote American anthropologists Christy and Jacqueline Turner in their book *Man Corn*. "Cannibalism has occurred everywhere at one time or another."

Going back at least until Paleolithic times, tribal cannibalism lingered until two centuries ago in some isolated South Pacific cultures, notably in Tonga and Fiji. Remnant cannibal cults survive today, and this book is an account of my quest to seek them out. For decades I have wandered the remote reaches of our planet, making journeys to the edges of the world, to be with tribes such as the Korubo of the Amazon and the Dani of New Guinea, tribes that appear to have changed little over thousands of years. Within a generation or two most of these cultures will disappear into the globalized monoculture, thrust into oblivion by the enormous appeal of Western ways once exposed to them.

It was on this search a decade ago that I encountered a tribe of treehouse-dwelling cannibals deep in a remote New Guinea rainforest on the much wilder Indonesian western side of the island. They call themselves the Korowai and they inhabit one of the world's most pestilent jungles, about one hundred miles inland from the Arafura Sea. It is not far from where Michael Rockefeller, son of Nelson, disappeared in headhunter and cannibal territory while collecting funerary totem poles for the Metropolitan Museum of Art almost five decades ago.

My first visit had been brief, as my Korowai guide would not take me beyond "the pacification line"—the line separating the Indonesian-controlled territory that was once under the influence of Dutch colonialism, and the area in which the Korowai practice a shocking form of ritual cannibalism. My guide knew of the tribesmen's threat to kill any outsider who dared enter their territory, and so he would not take me any further inland than his treehouse near a government settlement. This time I have gone beyond the pacification line in an effort to meet Korowai, who are among the last tribal cannibals on Earth, and to ask them why they eat each other.

I have visited several continents and large islands on the quest. In India the holy men of an ancient Hindu sect are fervent cannibals. They live at the cremation ghats by the sacred river Ganges in the holy city of Benares, and eat charred human flesh from the pyres as a demonstration of their saintliness. I went on to Tonga in the South Pacific where cannibalism is no more, but where until recently the kingdom's fierce giant warriors ate the flesh of their slain enemies. The Tongans even lined up Captain James Cook for the steaming oven when he sailed into their waters two and a half centuries ago.

In these three places the man-eating is and was condoned and even encouraged by their communities for specific, and, to them, honorable reasons. The Korowai kill and eat devil-men who haunt their treehouses, and they like the taste. The Tongan warriors humiliated slain foes by eating them, and they also liked the taste. The Hindu holy men use cannibalism as a way of reaching spiritual enlightenment, and even they are partial to the taste.

But I also traveled to Uganda, in eastern Africa, where cannibalism stemmed from a rebel leader more demonic than Hannibal Lecter. He has forced hundreds of boys and girls as young as six to eat human flesh after abducting them at night from their villages. The result is deep shame for a sickening purpose, to alienate the children forever from their families. The leader turns the boys into soldiers and the girls into sex slaves for his troops.

My last stop was Mexico where the sixteenth-century Aztecs have a deserved reputation for practicing mankind's most repulsive habit, even though they were a people of high artistic achievement. The Aztecs were enthusiastic cannibals, barbaric aesthetes. Central to their view of the world was the horrific need for daily human sacrifice and occasional cannibalism. In and around the great temple Tenochtitlán, at the center of what is now Mexico City, countless thousands of men, women, and even children were sacrificed to the Aztecs' bloodthirsty gods and then partially eaten.

Just thirty miles to the northeast is the mysterious two-thousand-year-old pyramid complex of Teotihuacan, where the genesis of much of Mesoamerican culture was sparked. I met a Mexican physical anthropologist who examined piles of human bones from fifteen-hundred-year-old garbage dumps there seeking to prove or disprove the traditional belief that the Teotihuacans were not cannibals. In diligently stripping away the flesh of cadavers in his laboratory with Aztec-style obsidian knives he has seen a marked similarity to butchers' cut marks on the ancient human bones.

There is more to my quest than just an adventurer's wanderings. In 1979, American anthropologist William Arens published a book, *The Man-eating Myth: Anthropology and Anthropophagy*, which immediately became highly influential in the often hostile tribal-and-ritual-cannibalism debate. Arens argued strongly that tribal and ritual anthropophagy—man-eating—had never existed: "I have been unable to uncover adequate documentation of cannibalism as a custom in any form for any society. Rumors, suspicions, fears and accusations abound, but no satisfactory first-hand accounts."

Arens gathered some convincing evidence for his case. The word "cannibal" comes from a corruption of the name of a tribe called the Carib, whom Columbus encountered during his epic voyage to America in 1492. An enemy tribe told Columbus that they were eaters of human flesh, a dubious proof. Columbus called them "caribales," which the Spanish later adapted to "canibales."

During Columbus's second voyage, in 1493, he supposedly found evidence of cannibalism, human bones in a hut according to the fleet doctor, Diego Alvarez Chance. Peter Martyr d'Anghera, the Spanish Royal Chronicler at the time of Columbus, and tutor in the court of Ferdinand and Isabella, had face-to-face access to the great explorer. He embellished the gruesome tale above by adding that there were also chunks of human flesh roasting on a spit, and a young boy's head hanging from a ceiling beam, still dripping blood.

Arens's refutation of such stories emboldened many anthropologists who refused to believe there was any evidence of tribal and ritual cannibalism. Typically, they claimed it was all a deception, lies put about by Western explorers, missionaries and colonizers to demonize the reputation of native peoples and so make it easier to exploit, enslave and convert them.

The bellicose Tolai of Papua New Guinea would scoff at Arens's disbelief. They were well aware of their cannibal ancestors' taste for human flesh.

The BBC reported in August 2007 that "the descendants of cannibals in Papua New Guinea (the Tolai), who killed and ate four Fijian missionaries in 1878, have said sorry for their forefathers' actions. They held a ceremony of reconciliation, attended by thousands, in the East New Britain province where the four died. The missionaries were part of a group of Methodist ministers and teachers who arrived in 1875 to spread Christianity."

In Mexico City, anthropologist Alejandro Terrazas Mata also scorned denials of tribal and ritual cannibalism. "William Arens is a nice old man, but he's not a scientist," he told me. He agreed with the Turners. "The evidence is overwhelming that peoples have practiced cannibalism all over the world through the ages. Even the medieval Europeans practiced a form of cannibalism when they ate pieces of human mummies to cure certain diseases."

It seemed incredible to me that there are cannibals in the twenty-first century who are not mentally twisted individuals, but well-adjusted and respected members of their communities, and so I sought to find them. I have not stuffed the following pages with stomach-turning accounts of human flesh eating. I have gone among the cannibals and observed their cultures to gain a greater understanding of who they are, and why they eat human flesh.

Sleeping with Cannibals

THE KOROWAI OF NEW GUINEA

CHAPTER 1

For almost three hours the Indonesian jet has been crossing the Banda Sea high above the clouds. We departed from Bali at 5 am, and the drone of the engines has lulled me into an uneasy slumber, until my neighbor, a soldier in camouflage fatigues, nudges me in the ribs. He points out the window to a turquoise coastline fringed with gardens of coral. Beaches and mangrove swamps give way to dense jungle that flows inland for several miles to be met by steep hills. Beyond the jungle rises an enormous spine of purple mountain, a towering massif crowned with jagged chunks of slate. The mountain is an omen; distant, elusive, dangerous. We are crossing the coastline of the world's second biggest island, a place once thronged with cannibals and headhunters.

"Papua," the soldier says in a soft voice that belies his hard dark eyes and grim smile, using the name for the province that forms the western half of New Guinea.

It is like returning home, because I came to New Guinea for the first time in 1961 as a youngster and have been back many times, lured by the seductive opportunity for adventure among the great island's fierce tribes. But this trip will be unique because I come in search of the last cannibal tribe on Earth.

Beyond the coast, New Guinea is a land of massive sheer-sloped mountains soaring as high as sixteen thousand feet, rammed together for hundreds of miles on end, and swathed in jungle. I keep glancing out of the window at the purple mountain, gripped by its majestic power. But banks of cloud drift in from the north-east and, with a suddenness that surprises me, the mountain disappears, leaving only the coastline and a thick cover of cloud that stretches from just above the jungle canopy to the top of the sky.

I have come to this distant place to revisit the tribe of cannibals known as the Korowai, who live in treehouses in one of the island's most far-off jungles. Ten years earlier, led by a guide, I had encountered the cannibals along a remote river one hundred miles in from the southern coast. But because it was too dangerous to go much further into their tribal territory, my guide would only take me a few hours' walk into the jungle to a Ko-rowai treehouse. I spoke with the clansmen there, but as they had been in contact with Dutch missionaries for several years, they were familiar with the horror outsiders feel about cannibalism. They would not show me any evidence that they had killed and eaten other men.

Nothing is holding me back this time. I am going to trek deep into Korowai territory, far upriver where there are clans who have never seen an outsider. For four decades I have roamed some of Earth's most remote places, seeking out people still living in what anthropologists call the Paleo-lithic era, but have never found a tribe that has not yet seen Europeans. I am perversely intrigued to discover whether the cannibals will welcome me with smiles or with arrows.

The Korowai tree-dwellers are among the most isolated people on Earth. They live in a country that was ruled by the Dutch for more than a century, but has been wrapped in the rough and often violent embrace of Indonesia since 1963. Their ferocity and the rugged terrain have warded off outside influences.

Most Korowai still live largely as they have for millennia, with little or no knowledge of the outside world, apart from the Citak headhunters who inhabit the rivers that sweep around the cannibals' jungles. When they were not fending off attacks by the Citak, who were ever eager to lop off heads, the Korowai have been constantly at war with each other, killing and eating with gusto male witches they call the *khakhua*.

An hour after crossing the coast, the jet descends over a wide lake dotted with straw-hut fishing villages and lands at Sentani, the airport of the pro-vincial capital, Jayapura, in the shadow of an emerald mountain. The jet drops onto a concrete strip that General Douglas MacArthur ordered laid down in 1944 as a vital stepping-stone in his push to dislodge the Japanese from the islands they occupied to the north.

The airstrip has a forlorn look with cracks in the tarmac and gardens of weeds thriving around its edges. The potholed road into Jayapura winds by green mountains with plunging slopes and coastal inlets where children splash in the bay, diving from straw huts perched on long poles over the water. Near the town, I begin to see the Indonesians, immigrants and carpetbaggers who flocked to Papua following its takeover.

The takeover happened more than four decades earlier, when, in a cynical Cold War romance, John F. Kennedy wooed the flamboyant Indonesian president Sukarno away from the Soviet Union and into the embrace of the United States by promising the dictator control of what was then Dutch New Guinea. Kennedy pressured the United Nations to transfer control from the Dutch to Indonesia. With a stab of the pen the mighty UN betrayed the one and a half million tribespeople from Biak in the far west to Merauke in the far south.

Most Papuans, like the Korowai, lived in the jungle hinterland, and had no idea that their land had been grabbed by a ruthless outsider in a grubby international deal. But the educated elite, who mostly lived by the coast, felt betrayed in their desire for full independence or amalgamation with their Melanesian cousins across the border in Papua New Guinea. They never wanted one colonial overlord replacing another.

The Indonesian army terrorized the new province into a reluctant acceptance of its new masters. Jakarta then accelerated plans to tame, exploit and populate its timber and mineral-rich frontier province, prompting fortune seekers to pour into the province from hopelessly overcrowded Java and other islands. Over three hundred thousand transmigrants settled in Papua, but they hugged the coastlines, rarely venturing into the jungles.

Since then, Jayapura has been transformed into a Javanese-style metropolis. In the main street, Jalan Ahmad Yani, the conquerors, Javanese soldiers in wraparound sunglasses and tight khaki uniforms that hug their slim taut bodies, strut along the crumbling sidewalks. Javanese shops and food stalls line the potholed streets, and Javanese bureaucrats occupy the desks in the government offices.

The food stalls are thronged with immigrants from across the archipelago, Javanese, Sumatrans, Buginese, Balinese, Macassans and many more ethnic groups, and they squat on the sidewalks or sit on stools as they slurp up bowls of pungent chili noodles. They threaten to overwhelm the locals, dark-skinned Melanesians with tightly coiled hair, who share little in common with the outsiders. (*Papua* is the Bahasa Indonesia word for "frizzy hair.")

Indonesia has targeted Papua's multi-billion-dollar natural resources, most destructively its lucrative stands of rainforest timber, among the world's largest. Of its one hundred million acres of virgin forest, almost seventy million have been licensed for logging. Vast forests have already been hacked out of existence. Even remote tribes like the cannibal Korowai, cut off from the rest of the world for thousands of years, will eventually come under threat from the relentless destruction of their land.

Some Papuan tribes are fighting back, mostly with bows and arrows against the army's assault rifles, in a brave but so far hopeless struggle to regain their land. This fight has prompted the Indonesian government to require any foreigner who wants to go inland to apply for a *surat jalan*, literally a "document to walk." This delays my journey to the Korowai by a day while the bureaucrats process my application in Jayapura.

The next day's flight takes me to the Grand Valley of Baliem. At five thousand feet above sea level, it is a fifty-mile-long stretch of fertile land that sits like the floor of a giant amphitheater surrounded by ten-thousand-feet-high walls of towering limestone. A coffee-hued river winds in tight loops along the valley floor. Dotted across the valley, clusters of straw huts with conical tops, like Chinese coolie hats, sit strategically on rises of land and by the foothills to blunt surprise attacks by enemy clans. Smoke spirals into the air from cooking fires, mingling with clouds that drift by at treetop level.

The jet touches down at a small hill town called Wamena. My nerves are jumpy as I cross the tarmac looking for my guide to the Korowai. It is not every day that you journey back to history's misty reaches to be with a tribe of cannibals.

Seeing my guide, Kornelius Kembaren, waiting at the airport gate, puts stuffing into my confidence. He is a daunting sight, a stocky north Sumatran tribesman who came to Papua sixteen years ago seeking adventure. As he strides across the tarmac, he looks just the man to lead me into the land of the cannibals. Clad in khaki shorts, polo shirt, long socks, and trekking boots, he has the unflinching gaze and hard, square jaw of a Marine drill sergeant. Kornelius first visited the Korowai in 1993, living in their tree-houses, and knows more about the tribe than most other outsiders.

With him is a short wiry young man, barefoot and clad in well-worn shorts and a T-shirt. After shaking my hand in welcome, Kornelius tells me that his companion, Boas, is a Korowai warrior who came to Wamena two years before to see what civilization looked like beyond his treehouse. He hitched a ride on a small plane, a charter flight, the only way to travel from Yaniruma, a settlement of about two hundred people at the edge of Ko-rowai territory. Boas has tried to return many times, but no one would take him. His father, according to Kornelius, is what the Korowai call a "fierce man," a war chief prone to violent outbursts.

Boas smiles. "My father became so angry because I hadn't come home that he's twice burned down our treehouse in frustration."

Because our destination is Yaniruma, Kornelius asks if we can take him along. When I agree, Boas's lips spread in a wide smile, but his eyes have a look of disbelief, as if he still cannot accept that at long last he is going home.

"Where's our chartered plane?" I ask Kornelius, expecting it to be wait-ing on the tarmac.

He shakes his head in disgust. "The airline just told me they've changed the schedule. We're departing tomorrow morning at eight."

"*Tida apah!*" I shrug. It's an Indonesian catch-all phrase, like mañana in Spanish, that shrugs off laziness, omission and sheer carelessness. "No wor-ries, mate," I add in Australian. "Let's go visit a Dani clan in the valley."

As a youngster I had sharpened my taste for adventure among such highland tribes. Shining with pig fat, sporting headdresses of golden birds of paradise plumes, flaunting crescent breastplates carved from mother-of-pearl and with their faces and bodies swathed in crimson and yellow ocher, the highland warriors are the grandees of New Guinea. As they stride the mountain passes and villages, they carry themselves as proudly as any me-dieval Spanish count. Any slight on their manhood, or any dispute involving women or pigs quickly brings out the duelling weapons: bows and arrows, spears and stone clubs. Then, hundreds of warriors face off against each other because clan fighting defines the meaning of life for highland men.

Apart from their vendettas, the warriors mired for centuries in the tribal demand for payback murder, the Dani men are distinguished by their custom of wearing only penis gourds known as *koteka*. Otherwise naked, the Dani flaunt giant mock penises made from gourds that rear up in a way that would startle even the most experienced paramour.

Many of the men are paired off and holding hands, but this does not necessarily mean that the two are lovers. It is the way men show friendship here. As they stroll along the crumbling sidewalks, careful I imagine not to get their *koteka* entangled, a bizarre thought pops into my mind. If two Dani men get into a fist-fight, down below do their *koteka* duel like fencing épées?

The road keeps to the valley floor, passing villages on either side that nudge the thoroughfare or nestle among the foothills. Above them soar the sheer slopes of the mountain walls, marbled by drifting cloud. The fertility of the land is amazing, the patchwork fields and forests fed by a rainfall of about seventy-five inches a year, and nourished by the hard work of the green-thumbed Dani.

The Landcruiser turns off the road and parks by a reed fence, the territorial boundary of a Dani clan. We stride along a dirt path meandering between a mosaic of fields, clumps of banana trees and rows of sweet potatoes, toward a village perched on a rise in the land.

The village hums with serenity, but nearby is a reminder that for thousands of years, until recently, war chants and battle cries echoed daily across this valley. Looming over the fields is a watchtower made from tall poles lashed together with vine. Crowning the tower is a platform with just enough room for a lookout.

A young warrior clad only in a *koteka* comes out to greet us with tight lips and cold eyes. His hair is coiled, soot-stained and soaked in pig fat. Kornelius gives him the equivalent of five dollars, and, having offered a fleeting grin in return, he leads us up to a log stockade that guards the entrance to the village.

Crossing the knee-high step is like stepping back ten thousand years. Wispy clouds float by, a few yards above several mushroom-shaped straw huts clustered around one big hut in the middle of a cleared space. An elderly woman sits by one of the smaller huts. She is smeared in white clay, and I notice that she has several joints missing on some fingers. I know from my time in the highlands that whenever a relative died in a battle, a girl had to give part of her finger to placate his spirit. The Dani world is full of vengeful ghosts who must be appeased with bloody rituals for the clan to remain prosperous and powerful.

Not every New Guinea tribe adheres to this custom. I once visited this same village with a Korowai warrior named Agoos. When he saw women with mutilated fingers at the Dani village he snorted, "What a waste. How can your woman gather and cook food and raise children with her fingers chopped off?"

As with that earlier visit, this time a young warrior leads us into the men's hut, dark and smoky, where all the men and adolescent boys sleep huddled each night. Bows, arrows, war axes and spears are stacked against the walls ready for the call to arms. Dry grass is spread across the floor. A flame flickers in the fireplace scooped out of the floor, but there is no chimney and the smoke is trapped inside. My eyes begin to sting and water, and it is hard to see through the shadowy gusts of smoke.

By the wall squats the oldest man I have ever seen, a living mummy, his skin shrunk to the bone, the flesh on his face barely clinging to the bones, his long bony arms clasped about his ankles. Close-up I see that he *is* a mummy, his skin smoked the color of ebony. A band of feathers perches on his skullcap. "He's my ancestor, and he's over three hundred years old," says the warrior. "He was our greatest war chief, and led our clan to victory in many great battles. That's why we still honor him."

On the earlier visit, about a dozen men had come in from the fields and stood around a fire to ward off the chilly mountain air, their long skinny arms clasped about their backs. The clan's war chief, named Jaja, peered at Agoos, the Korowai, as if he was trying to fathom from where Agoos came. At first glance, in his tattered shorts and T-shirt, he must have looked like many of the outsiders who drifted about Wamena's dusty streets, men come from tribes all over the province to do the servile work of their Indonesian masters, tasks too groveling for the proud Dani. But although the Dani warriors towered over Agoos, there seemed something else about him that caught the chief's attention.

"Where is this man from?" Jaja asked as he pointed at Agoos.

"He's from the south, a Korowai, a cannibal," my guide explained.

The chief pulled back as if he had been slapped in the face. "He eats humans," he said softly. Fear throbbed in the eyes of this hero of dozens of battles, this man raised to be a ruthless war chief in a warrior culture. The word spread among the Dani. "Cannibal," they murmured in dread as they backed away from Agoos.

I had not expected the Dani to show alarm before a lone outsider, especially within the safety of their own stockade. But the human fear of cannibalism was so vivid, seared into their minds, that even the bravest of the brave Dani cringed before a man known to be an eater of humans. Agoos let slip a smile of triumph as he watched the war chief and the other Dani warriors retreat, but kept silent. He too was raised to fight and kill other men, and he must have enjoyed seeing the warriors of another clan, men much bigger than he, tremble before the power of his presence. It was this commanding blend of charisma, machismo and bravado that another warrior society, the Polynesians, called a warrior's *mana*.

☠

It is mid-afternoon when Kornelius drops me at a small hotel at the edge of town. "I'll see you at dinner tonight," he murmurs, and then speeds off.

I am happy he wants to have dinner with me so we can get to know each other. He has not once smiled, and I wonder whether he will loosen up at the dinner, or whether I will have to wait until we enter the Korowai jungle.

You do not go to New Guinea to indulge in gourmet meals. Dinner at the hotel, the best in Wamena, is a plate of greasy fried rice washed down by a brand of Indonesian beer that tastes as if it were brewed from spiders' eggs. Kornelius, being a Muslim, drinks bottled water and eats beef kebab, even though the waitress tells him that there is no pork in the *nasi goreng*.

I like his caution, a contrast to my own preference for walking along the edge of every cliff I encounter in life. One Errol Flynn on such a trip is more than enough, and Kornelius has the task of bringing us to the cannibals and then getting us out safely. Still, I hope he will soften as we get to know each other better.

I ask how he ended up in Indonesian New Guinea. He settles his elbows on the table and fixes me with a steady gaze. "By choice. My father was a forest worker, and from him I learned to love the jungle and tribal people. When I graduated from university, I came here searching for clans still living in the Stone Age. I spent three years with the Asmat (the tribe that probably killed Michael Rockefeller), but they had become too civilized. I heard about the Korowai, and so I flew to Yaniruma, made friends, and within weeks I was living among them. I've spent months at a time in the treehouses."

"And now you're a friend of cannibals."

His lips form a feeble smile, but his eyes do not soften. "Sometimes, your friends choose you, not you them."

"I've read that tribal cannibals call human meat 'long pig' because the taste is similar."

Kornelius shakes his head. "I've never tasted pig, but the Korowai say it's nothing like pork."

"Then, what does it taste like?"

"It's better you ask them when you get there."

His eyes dim as he lowers his head, and I notice that his throat is pulsing, as if he wants to say something, but a force deep within him is preventing the words from forming. After a few moments of this inner pain, he says softly, "I ate human flesh once."

I shake my head in disbelief. "You're joking."

Kornelius holds his hands up apologetically. "No, it's true. About ten years ago, a Korowai man in Yafufla, a village we'll pass through, came to me one night and said that if I wanted to be trusted by the Korowai then I'd have to prove my good faith by eating human flesh. If I didn't, he warned, then I might just as well go away and never come back because no Korowai would want to be my friend."

He sighs, as if the memory of this gruesome test is still fresh in his mind. "I spent hours thinking about it, well into the night. And then I decided that unless I ate it, I'd never get to know the Korowai customs and their language. So, I sent a message to the man and he sent me back a chunk of meat wrapped in sago leaf, part of the thigh of a man the clan had killed a few days earlier because they believed he was a sorcerer. The meat was a bit tough, but the taste was good. I've never eaten human flesh since, and thankfully, they've never again put me to the test."

"Do you regret what you did?"

He sighs once more, and turns his eyes to his hands folded on the table. "I've never asked myself that question, and I never will."

Kornelius looks up and gazes directly at me. "What would you do?" His tone tips uncertainly between a plea for understanding and an accusation.

My stomach churns at the thought. "I'd refuse. I'd never eat human flesh."

"What if they threatened to kill you if you didn't?"

Doubt grips my tongue for a few moments. "I'd still never eat human flesh, whatever the consequences."

Kornelius shakes his head knowingly as if I were a simpleton, or a child believing in the absolute certainty of life's steady path after only experiencing just a little of its joys and terrors. "It must be nice to be so sure about everything. That's a city person talking. The worst you risk every day is being knocked over by a car. Out in the jungle where we're going, you could be faced with choices of life and death far removed from your experience. And then, believe me, sitting here in this hotel, and not under any threat, you have no idea of what you'd do out there. It's a different planet."

I shrug, not knowing how to answer this, and wave him away into the night.

CHAPTER 2

That night, flat on my back in a sparsely furnished hotel room, with an overhead fan lazily slapping the torpid air, I trawl through the controversial question of so-called institutionalized cannibalism, whether it exists or whether it is an historical hoax. The doubters' guru, anthropologist Willam Arens, stirred the pot in 1979 with his book *The Man-eating Myth: Anthropology and Anthropophagy.* He challenged, "I am dubious about the actual existence of this act as an accepted practice for any time or place." He contended that such cannibalism was "unobserved and undocumented."

At about the same time that Arens was casting doubt on the existence of tribal cannibals, Dutch pastors from the Mission of the Reformed Churches in the Netherlands were risking their lives by journeying to the Korowai. For two decades they suffered much hardship, living in rough conditions, as they studied the Korowai language and documented their culture, which had as a core ingredient ritual cannibalism, the killing and eating of village men branded *khakhua*, or witch-men.

The Dutch pastors were no culture destroyers. One of the missionaries, Gerrit van Enk and linguist Lourens De Vries, cowrote the only book on the tribe, *The Korowai of Irian Jaya* published in 1997 by the venerable Oxford University Press. Van Enk told me, "We went to the Korowai to convert them to Christianity, but when we got there realized we'd destroy their unique culture if we did that. So, we gave up our plans to convert them, and remained for many years as observers, documenting their culture which was so radically different to ours, and so ancient, while attempting to influence them as little as possible."

Unwilling or unable to tramp into the dangerous Korowai jungles to see for themselves, the doubters disputed this testimony with no evidence except their own prejudice. Arens wrote in 1998 that, "I continue to aver not only that the Caribs, Aztecs, Pacific Islanders and various African, native American and New Guinea tribes have been exoticized, but also—and equally important—that Western culture has congratulated itself for putting a stop to the cultural excess through colonial 'pacification' and introducing Christianity to once-benighted natives."

Arens's eagle eye roamed wide in geography and history in search of fake cannibals and he singled out New Guinea, an island infamous for centuries among Westerners for its cannibal tribes. (In 1998, the Duke of Edinburgh congratulated an Australian student for successfully traversing the swampy and mountainous Kokoda Trail in Papua New Guinea. "You managed not to get eaten then," he joked.)

Besides van Enk and De Vries, Arens could have found numerous first-hand witness accounts of cannibal feasts there. My first ever job was as a trainee colonial official in Papua New Guinea. I flew from my home in Sydney to the capital Port Moresby on my eighteenth birthday, and among my friends there were patrol officers, or kiaps. These brave men risked their lives to take the rule of law and the liberating concept of modern Western-style justice to remote tribes who had been living under a simple and brutal code of kill-or-be-killed for millennia.

The patrol officers filed meticulous reports of their contacts with people still living in the Stone Age, and among these is considerable evidence of payback murder and ritual cannibalism among the tribes. In 1920, a patrol officer turned anthropologist named Ernest Chinnery wrote of cannibal raids in the Gulf province, not too far across the border from Korowai territory. Chinnery lectured about his finds to the Royal Geographical Society in London, and in 1920 won its Cuthbert Peek award for his anthropological work among the Papuans.

"The people had been classified as Papuan, and all, except those under control, practice headhunting and cannibalism," he noted in his report to the colonial authorities almost a century ago. "Before a house can be occupied or canoe launched it is the custom to sprinkle the building or boat with human blood . . . the heads of people are slain and collected . . . bodies are cut up, cooked in various ways and eaten." In one village Chinnery visited, he reported, "The people were living in one large house, which was entered by my party at dawn while the Moreri were eating the bodies of Irumuku natives they had killed."

Like Chinnery, the other patrol officers mostly wrote their reports in a matter of fact style, even their accounts of cannibalism, and, in contrast to the slander of doubting anthropologists, usually resisted the temptation to make judgments on the locals' sense of morality or the notion that the kiap with his Western culture was in any way superior to these people.

Just two years before I made my first trip to Papua New Guinea, seventeen warriors from the Mianmin tribe on the Sepik River, about three hundred miles northeast of Korowai territory, formed a raiding party and attacked a communal hut inhabited by the Sowana tribe. They killed six men and abducted their wives. John Mater, a patrol officer, investigated the attack at the remote location, and wrote in his official report that the raiding party cut up the bodies with bamboo knives. "The Mianmin left the heads and entrails of the victims and carried the rest of the bodies away to be eaten." The raiding party carved out the livers as snacks, and after several days, when they arrived home, "the remains of the bodies were cooked with taro and eaten."

At their trial a year later, in which the men admitted guilt, the judge noted that "Apparently they have to rely on raids of this kind to obtain wives for their young men, and that the killing, cutting up and eating of the women's husbands appears to be accepted by the women as something inevitable and final."

The Amazon was another hotbed of tribal cannibalism. Beth Conklin, an American anthropologist at Vanderbilt University, studied a tribe there whose members were until recently cannibals. She said: "We assume that cannibalism is always an aggressive, barbaric and degrading act. But this is a serious over-simplification, one that has kept us from realizing that cannibalism can have positive meanings and motives that are not that far from our own experience."

The cannibal doubters would probably wrinkle their noses and snort disbelief. But as with the Dutch missionaries and the Korowai, Conklin did the hard yards in the Amazon, journeying to the Wari jungles and shrugging off the tribulations to live with them for nineteen months, between 1985 and 1987. The Wari were cannibals until missionaries and government pacification teams in the 1960s stopped them from eating human flesh. Older tribespeople confirmed to her their cannibalism, and missionaries and government officials described to her how they had witnessed cannibalism among the Wari in the 1950s and 1960s. Conklin returned to the Wari three more times, the latest visit in 2000, to reconfirm her findings.

The Wari described to Conklin how their cannibalism came in two forms. "Eating enemies was an intentional expression of anger and disdain for the enemy," she said. "But at funerals when they consumed members of their own group who died naturally, it was done out of affection and respect for the dead person and as a way to help survivors cope with their grief." This mortuary cannibalism "helped mourners emotionally detach from memories of the dead, which in turn helped them deal with the loss of a loved one. For the dying, being incorporated into fellow tribesmembers' bodies was far more appealing than being left alone to rot in the dirty, wet, cold and polluted ground."

She went on to say that the Wari "may have understood ways that made the destruction and transformation of the body through cannibalism seem to be the best, most respectful, most loving way to deal with the death of someone you care about."

The work of Jens Bjerre, a distinguished senior fellow of the Royal Geographical Society of London, took place closer to my own early wanderings. His book *The Last Cannibals*, published in 1956, tells of encountering cannibalism among the fierce Kukukuku, who live in the rugged southern foothills of Papua New Guinea. I was based not far away in 1962, and at that time they were still regarded as the most warlike of any tribe in the Australian colony, the eastern half of the great island. Short, stocky and truculent, the bare-chested men wore bark skirts tied at the waist and carried their bows and arrows with them wherever they went, even to their outside hole-in-the-ground toilets.

I only ever saw the Kukukuku once, when a band of warriors was brought to Mount Hagen, the western highlands' provincial capital where I lived, for the annual "sing-sing" of the tribes. The field set aside for the sing-sing was thronged with several thousand highland warriors, tall strong men who made the hills shake with their bellicose chanting during spear dances. Their vividly painted faces gave them the appearance of Technicolor demons. But when the highlanders saw the ten Kukukuku warriors approaching, there was a sudden hush and they warily shuffled back en masse to allow the outsiders to pass through the throng without challenge. They well knew the Kukukuku's fearful reputation, gained by centuries of bloody raiding, and I could see fear in the highlanders' eyes as the little men strode disdainfully past them.

Bjerre bravely trekked into the Kukukuku hills with an Australian government patrol while I was still in high school. He observed:

> When a party of Kukukuku warriors takes an enemy prisoner, either in combat or by abduction, they tie the captive to a thin tree-trunk and bring him horizontally back to the village. So that the prisoner shall not escape, they then break his legs with a blow of the club, bind him to a tree, and adorn him with shells and feathers in preparation for the forthcoming orgy.
>
> Fresh vegetables are brought in from the fields and a big hole is dug in the ground for an oven. As a rule, the children are allowed to "play" with the prisoner; that is to say, to use him as a target, and finally stone him to death. This process is designed to harden the children and teach them to kill with rapture.
>
> When the prisoner has been killed, his arms and legs are cut off with a bamboo knife. The meat is then cut up into small pieces, wrapped in bark, and cooked, together with the vegetables, in the oven in the ground. Men, women and children all take part in the ensuing orgy, usually to the accompaniment of dances and jubilant songs. Only enemies are eaten. If the victim is a young strong warrior, the muscular parts of his body are given to the village boys so that they can absorb the dead man's power and valor.
>
> Jack (an Australian patrol officer) told me that, six months ago, two men had been eaten in a village, Jagentsaga, not far away; and that a month ago he had, by chance, found the hand of a man who had been eaten shortly beforehand. The rest of him had been hidden in the jungle. "They know," he said, "that we will punish them for cannibalism, so they do everything to conceal it now. But it still occurs and probably will do so for a long time."

Arens and his supporters seem to have shrugged off the wealth of such easily available evidence of tribal cannibalism. Enough of this musing for tonight, I think to myself, or I might suffer nightmares and I will need to be up early for our departure. Within days in a remote jungle I will come face to face with men whom Kornelius says truly are tribal cannibals.

CHAPTER 3

At the airport the next morning Kornelius nods a greeting to me but says nothing. It makes me wonder whether confiding in me about his brief acquaintance with eating human flesh embarrasses him. By his side is Boas, carrying his few possessions in a tote bag. He has donned a daisy-yellow bonnet, perhaps a reminder to him of the "civilized" ways of the world outside his Korowai jungle.

With a chilly wind blowing in our face we stride across the tarmac and board our chartered Twin Otter, a sturdy little workhorse whose short take-off and landing ability will get us into our destination, Yaniruma. As we fly down the Grand Valley, the sheer walls of the high mountains on either side angle in at the valley's green fields. The highlands are the most densely populated part of rural Papua, and huts dot the valley and foothills that nudge the mountain slopes.

Slicing through this Stone Age metropolis is the Baliem River, picking up speed as it rushes to surge through a gap in the mountain chain, the Baliem Gorge. Here, the mountains suddenly part, freeing the river to tumble down to the jungle far below in rainbow-misted cataracts.

We emerge from this giant's bolthole above a lowland jungle that flows almost unbroken to the southern coast. As we fly above the rainforest's never-ending tangle, Kornelius shows me our destination on a map spread across his legs. Surrounding the spidery lines marking the lowland rivers are thousands of square miles of jungle blanked out and shaded green. Forbidding, untamed. Dutch missionaries who came to convert the Korowai called it the hell in the south. Yet, this hell on earth was like a magnet to Christian missionaries, mostly from Holland, and then much later from the United States, who came with the intention of battling Satan in such a devilish place.

They were determined to convert these "demonic people," cannibals and headhunters. The missionaries spread out along the relatively accessible coastlines beginning in the mid-1800's, eventually to be constrained by a curious pact enforced by the Dutch colonial officials called the Boundary Line of 1912. It mirrored the geographic separation of religion in their homeland, with Protestant missionaries keeping religiously to the north and Catholics confining their missions to the south.

Toward the end of the ninety-minute flight, we fly low following the snaking Ndeiram Kabur River. In the jungle below, Boas spots his father's treehouse. It seems impossibly high, like the nest of a giant bird. He trembles with emotion and, as he hugs me, tears trickle down his cheeks. Over and over he utters a Korowai word, *manop*, meaning "good." I soon come to know that it is one of the tribe's most common words.

"There are about four thousand Korowai, and their treehouse clearings are sparsely dispersed over several hundred square miles of swampy jungle," Kornelius tells me over the roar of the Twin Otter's engines as he points down at another towering treehouse dominating a clearing in the jungle about the size of two football fields. "Each clearing has one or two treehouses inhabited by up to twenty people, and most of the adults are cannibals."

At Yaniruma we thump down on a dirt strip Dutch missionaries had carved out of the jungle near the Ndeiram Kabur. A line of stilt huts stretches back from the airstrip's fringe to a murky stream that leads to the river. Waiting by the strip are dozens of women and children, Korowai, clad in ragged dresses and shorts, who look wide-eyed at me as I step from the plane. "They've walked for a day or more to see you," Kornelius says. "They call white people *laleo*, or ghost-demon, and it's a long time since the Dutch missionaries lived here."

"Are they cannibals?"

Kornelius nods. "Most of the women have eaten human flesh, but the children are forbidden to touch it until they're about fifteen."

I want to ask Kornelius if the children look forward to the day when they come of age and are invited to the cannibal feast, but hold my tongue, afraid to hear the answer.

Boas is moving from friend to friend gathered by the airstrip, enveloping each in a hearty hug. "And what about Boas?"

"Of course. Almost all Korowai adults I've met are cannibals."

I point to a group of short, stocky men who are unloading our luggage and supplies from the plane. "Are you comfortable with a bunch of cannibals taking us deep into the jungle?"

For Kornelius, the answer is simple. "Yes. Understand this. If you were born a Korowai, you'd be a cannibal. When you realize that, being friends with them is easy." With most of them anyway—one spooky looking tribesman named Kili-Kili has remained aloof. "He's killed and eaten thirty *khakhua*," Kornelius says. "When we get to Yafufla I'll hire him as a porter."

☠

The first outsider to dare come here, in 1979, was a missionary from the Dutch Mission of the Reformed Churches, the Reverend Johannes Veldhuizen, who established Yaniruma as a base from which to bring the word of God to the Korowai. Veldhuizen was appalled by their cannibalism, and at first he thought he would try to stop them from eating human flesh. But once he came to know and be impressed by Korowai culture, he abandoned plans to convert the tribe. I had spoken to him by phone in the Netherlands. "A very powerful mountain god warned the Korowai that their world would be destroyed by an earthquake if outsiders came into their land to change their customs. So, we went as guests rather than as conquerors, and never put any pressure on the Korowai to change their ways."

As a result, few Korowai have ever become Christians, and the Dutch missionaries are long gone. In the early 1990s the Indonesian government refused to renew the work permits of Veldhuizen and his colleagues.

That does not mean Indonesia has given up trying to convert the Korowai. It is using the settlement at Yaniruma to lure the Korowai into leaving their jungles and their "demonic" customs behind, but the Jakarta government's motive is in the service of Mammon rather than God. The vast stretches of the Korowai's dense rainforest, which harbor billions of dollars worth of hardwood timber, have been carved up among prominent

Indonesian politicians and generals into lucrative timber concessions, and they want the Korowai out of the way.

So far, only a few hundred Korowai have taken the bait—they are given a hut and land at Yaniruma and a handful of other settlements along the river. Most of the tribespeople still prefer their life in the jungle. The government exerts no pressure on them to move for the time being because they can exploit timber stands much closer to the coast, and the lumbermen will only move on the Korowai rainforests when those others have been fully exploited.

Nothing seems to have changed at Yaniruma in the decade since I was last here. It is still a sleepy little backwater with its tumbledown stilt huts engulfed by weeds. The heat at mid-morning is mind-numbing, and most of the settlement's inhabitants are lounging on rickety verandas, chatting or sleeping. Its gentrified Korowai discarded jungle clothing, and are now clad in the same kind of torn and tattered grubby shorts, T-shirts and dresses I have seen many times in fringe tribal settlements along the Amazon and the Congo, and now the Ndeiram Kabur.

"We're not going to linger here," Kornelius says as he hoists his daypack on his shoulders. Boas has already grabbed a large sack of rice, now balanced on his head. "I've decided not to go home yet," he tells me. "I'm coming upriver with you."

"But your father must be very eager to see you."

"I'm sending word by my brother who lives in Yaniruma that I'm back, and that I'll be home in a couple of weeks." I notice he is gripping a bow and several barbed arrows. "You helped me when it looked like I'd never be able to leave Wamena. I know the clans upriver, they might try to kill you because you're a *laleo*, and I want to defend you."

Kornelius knows the Korowai well, but even he has never dared go far up the Ndeiram Kabur River, our destination. As we sit by the airstrip, watching Korowai men being signed on as our porters, he warns me of our journey's dangers. "There are clans upriver who threaten to kill outsiders daring to enter their territory," he says. "They especially fear and hate those like you with white skins, the *laleo*, even though none have ever seen white people. They were warned of your presence beyond their jungles by tales their shamans have told around the campfires for as long as the Korowai can remember."

As the pilot hurls the Twin Otter back into the sky, Boas joins our porters as they begin trudging toward the nearby jungle in single file, carrying our packs and supplies. Most, like Boas, also carry bows and arrows.

"Why?" I ask Kornelius.

"In case we're attacked."

The Reverend Gerrit van Enk gave me the clue when I phoned him at his home in Holland. When I told him about my plan to journey several days upriver, he sighed with envy. "I never got that far because it was beyond the pacification line. Be careful up there because the Korowai are a volatile people, and can explode into violence against you at any time, sometimes without you knowing the reason why you upset them."

As we trudge through the settlement, I expect to be challenged by an Indonesian policeman, demanding to see the *surat jalan* I carry. No challenge comes. "The nearest police post is at Senggo several days back along the river," Kornelius explains. "Occasionally a medical worker or official comes here for a few days, but is too afraid of the cannibals to go deep into Korowai territory. The government mostly leaves the Stone Age Korowai alone."

☠

Entering the Korowai rainforest is like stepping into a giant watery cave humming with malice. One moment, with the bright sun overhead, I breathe easily, but as the porters push through the undergrowth, towering trees abruptly close in over us, blocking the sky. The heat is stifling and the air drips with humidity, clutching at my throat. I peer at the shadowy world ahead, plunged into a verdant gloom by the dense weave of the canopy. It is the haunt of giant spiders, killer snakes, murderous microbes, warring cannibals.

High in the canopy parrots screech a warning as I follow the porters along a barely-visible track winding through rain-soaked trees and primeval palms. My shirt clings to my back and I take frequent swigs at my water bottle. We are constantly dive-bombed by squadrons of large black-striped mosquitoes. I worry that some carry malarial parasites. Other mosquitoes bear the microscopic worms that cause the terrible affliction of elephantiasis that swells the legs and testicles to giant size. Only repeated dousings of insect repellent keep the pests at bay.

A torrential downpour hits without warning, the rain spearing through gaps in the canopy, but we keep walking despite the drenching. The annual rainfall in this flooded jungle is well beyond two hundred inches a year, making it one of the Earth's wettest places, and the forest floor swims with murky swamp water and pools of mud.

The local Korowai have laid lines of logs along the surface of the mud, and the barefoot porters cross it with ease. I, however, struggle. Desperately trying to balance as I edge along each log, time and again I slip, stumble and fall into the sometimes waist-deep mud, bruising and scratching my legs and arms.

More worrying are the slippery solitary logs, up to ten yards long, slung across the many dips in the land. While inching across like a tightrope walker, I wonder how the porters would get me out of the jungle were I to fall and break a leg. Again and again I mutter to myself, "What the hell am I doing here?" but I know the answer. To witness people still living as our ancestors might have done ten thousand years ago makes bearable the pain, misery and risks of this extreme trek.

Boas sees my difficulty and strides back along the line of porters to walk directly in front of me. When he encounters a tree fallen across the track, he thrusts an arm back to grab my hand and help me over it. When we come to a line of logs thrown across the mud, he walks by my side, ready to grab me if I look like I will fall. When we are confronted by one of the logs positioned across a hazardous drop, he goes first, gripping my hand as, step by step, he steers me to safety.

Hour melts into hour as we push deeper into this nightmarish place, stopping a few minutes every hour to give the porters a rest. With night near, my heart surges with relief when the jungle gloom dissolves as shafts of silvery light stab through the trees ahead, signaling a clearing. "It's Manggel, another village set up by Dutch missionaries for Korowai to lure them away from their treehouses," Kornelius says. "We'll stay the night here."

Curious Korowai children with beads about their necks come running to point and giggle at the *laleo* as I stagger into the village, a huddle of stilt straw huts overlooking the river. There are no old people here. "The Korowai have hardly any medicine to combat the jungle diseases or cure battle wounds, and so the death rate is high," says Kornelius. Middle age is a rarity, and few children ever know their grandparents.

Misty-eyed Westerners who have never been within thousands of miles of tribal people are often seduced by the eighteenth-century philosopher Jean-Jacques Rousseau's romantic image of the noble savage, living in peace and harmony with nature. But life in the Korowai jungle is pure Hobbesian, often short and usually brutal. The children die the easiest, and many do not survive their first year. That is why the children are not named until they are about eighteen months old. They then run a life-long gauntlet of killer tropical diseases and accidents without any medical help, and as adults risk the sudden thud of an arrowhead or the fatal swing of a stone ax.

Van Enk, drawing on years of experience with the Korowai, wrote that "interclan conflicts, the *khakhua* complex and diseases like tropical malaria, tuberculosis, elephantiasis and severe anaemia constantly engender victims."

The Korowai have no knowledge of the deadly microbes and germs that infest their jungles, and so, using Stone Age logic, believe that any mysterious death must be caused magically by a *khakhua*, a witch-man. After dinner Boas sits next to me cross-legged on the hut's thatched floor, his dark eyes reflecting the gleam from my torch, our only source of light. Using Kornelius as translator, he explains why Korowai, for as far as the tribe's memory goes back, have hunted down and eaten *khakhua*, monsters from the world of the supernatural that have taken on the form of men.

Until a person is dying, the Korowai are unaware that a *khakhua* is among them, using, as camouflage, a close relative or friend's living body. "The *khakhua* eats the victim's insides while he sleeps, replacing them with fireplace ash so that he doesn't know he's being eaten," Boas explains. "The *khakhua* finally kills the person by shooting a magical arrow into his heart."

The death spurs the victim's male relatives and friends to hunt down and eat the *khakhua*. "Usually, the victim whispers to his relatives the name of the man he knows is the *khakhua* that is killing him," Boas explains. "He may be from the same or another treehouse."

I ask Boas whether the Korowai kill and eat criminals, or the bodies of enemies they have killed in battle. "Of course not," he replies with a look of surprise. "We don't eat humans, we only eat *khakhua*."

The killing and eating of *khakhua* has fallen off in the settlements, but is still practiced out in the jungles. "A *khakhua* is killed and eaten about once a year in and around one of the settlements such as Yaniruma, but deeper in Korowai territory, many *khakhua* are murdered and eaten each year," Kornelius tells me, citing information he has gained in his frequent talks

with Korowai in settlements and in treehouse clusters. Whenever a Korowai man dies from disease there is terror in the treehouses nearby, with
most men dreading that the dying person will name one of them.

Rupert Stasch, an anthropologist at Reed College in Portland, Oregon,
lived among the Korowai for eighteen months, studying their customs, and
in an article for an Australian journal, *Oceania*, wrote that fear of the Indonesian police has prompted the fall-off. He told of a Yaniruma man a decade
ago killing his sister's husband for being a *khakhua*. When this was reported
the police flew in and arrested the killer and two accomplices. They took
them back to their base. "The police rolled them around in barrels, made
them stand overnight in a leech-infested pond, and forced them to eat to-
bacco, chili peppers, animal faeces and unripe papaya. Their ordeals in the
distant government seat have been widely recounted ever since."

☠

Nevertheless, the fear of being named a *khakhua* remains strong even
in the Indonesian-sponsored settlements. Ten years earlier I had been at
Mbasman, a settlement like Manggel, inhabited by Korowai families who
had moved from the jungle to take up the government's offer of a hut and
land. I arrived at night with my guide from Wamena, Agung, and we had
slept in the hut of the tough-faced, middle-aged village headman, Barnabas.

Soon after sunrise, angry voices had woken me. Agung squatted on the
porch of the hut surrounded by a dozen sullen-faced men clad in ragged
shorts. He pointed to the men. "We need six or seven porters and a guide
to get us to a Korowai treehouse I've been to once before, a few hours walk
through the jungle from here, but they refuse to go. They say the jungle
Korowai will kill and eat them if they enter the jungle. A man has just died,
and his clan is searching for the sorcerer who killed him."

I offered double the usual porter's fee, then tripled and quadrupled
it, but they would not budge. Fight fire with fire, I decided, and gripped
Agung by the shoulder. "Tell the Korowai that I'm a powerful shaman from
a far-off land, and I'll use my spells to protect them in the jungle."

On the porch, my mind raced back to the days when I was an altar boy.
"Please have the women, children and men come here," I asked Barnabas.

More than one hundred Korowai drifted from their huts to gather in
front of us. I gazed down at them and spread my arms like a priest, re-

mained silent for a few moments to heighten the suspense, and then looked up at the heavens with beseeching eyes.

Back in the present, Kornelius shakes his head. "That was a mistake because the Korowai don't believe in heaven or hell, with God high above and the devil way down below. They have three concentric circles of life with the living inhabiting the innermost circle, the dead existing in the middle circle and a vast stretch of water with huge fish in the third." I shrug and go on with the story.

Facing the Korowai I began chanting the Mass in Latin. "*In Nomine Patris et Filii, et Spiritus Sancti Amen*," I shouted in a holy-roller voice. "*Introibo ad altare Dei*."

A murmur rippled through the Korowai. I might well have been chanting in English for all it mattered, but the sonorous cadences of Latin gave me courage. The novelty effect, however, was short-lived, and the Korowai began muttering among themselves or walking away.

I shouted louder, rolling my *r*'s, making signs of the cross, holding my hands to the heavens, but saw that I was losing them. After finishing the *Confiteor Dei* I gave up. Agung glanced at me sympathetically. "It was a nice try. I could tell by what they were saying that they believe you have magical power, but it's only strong in your own land. We'll have to return today to Wamena."

Shortly afterward, as I was packing my bag, Agung called me back onto the porch. "Let's go," he cried boyishly, pointing at four men and a woman heading in single file toward the jungle with our supplies. "Barnabas forced them to take us to the tree people."

Our guide was a middle-aged woman who had cast aside her dress and donned Korowai jungle garb, a grass skirt a bit longer than a hand span. *Terima kasih*, I told her in Bahasa Indonesia. Thank you. She looked at me with dark eyes that throbbed with hatred and spat in my face.

Boas claps his hands in delight at my story and laughs. "She wasn't afraid she'd be killed and eaten as a *khakhua*, because a *khakhua* can only be a male," he explains. "Her husband must have been one of the porters, and you were putting him at risk."

CHAPTER 4

We trek from soon after dawn until just before dusk to reach the next settlement, Yafufla, another line of stilt huts, set up by the Ndeiram Kabur to lure Korowai from the jungle. They were largely unsuccessful. "After here, we'll only see treehouses," Kornelius promises.

After a dinner of river fish and rice spiced with burning hot sambal paste, Kornelius takes me to an open hut overlooking the river. Two men approach through the gloom, one in shorts, the other the first Korowai in traditional garb that I have seen on this trip. He's naked save for a necklace of pigs' teeth and a leaf wrapped about the tip of his penis which has been thrust back into his scrotum. "That's Kili-Kili," Kornelius whispers.

Kili-Kili carries a bow and barbed arrows, and is the scariest-looking human I have ever encountered. His dark eyes are empty of expression, his lips are drawn in a perpetual grimace, and he walks as if he were a shadow, without sound, slipping from space to space with little warning of intent. He is a mass killer whose humanity seems to have been almost entirely sucked out of him by the terrible things he has done. A Stone Age Terminator.

The other man is his brother, Bailom. He pulls from a bag a skull with a jagged hole marring the forehead. "It's Bunop, the most recent *khakhua* he killed," Kornelius says. "Bailom used a stone ax to split the skull open to get at the brains." His eyes dim with sadness. "Bunop was one of my best porters, a cheerful young man."

Bailom passes the skull to me. I do not want to touch this gruesome relic, but am afraid of offending the killer. My blood chills at the feel of naked bone. The fire's reflection flickers eerily on the brothers' faces and the skull as Bailom tells me why, two years before, he murdered Bunop. "Just before my cousin died, he told me that Bunop was a *khakhua* and was eating him from the inside. So, we caught him, tied him up, and took him to a stream where we shot arrows into him." Bunop had screamed for mercy, protesting that he was not a *khakhua*. Bailom was merciless. "My cousin was close to death when he told me, and would not lie."

At the stream, Bailom chopped off the *khakhua's* head with a stone ax. As he held it in the air and turned it away from the body, the others dismembered Bunop while chanting. Bailom shows me, with his hand making chopping movements, how it was done. "We cut out his intestines and broke open the rib cage, chopped off the right arm attached to the right rib cage, the left arm and left rib cage, and then both legs."

The bloodied body parts were individually wrapped in banana leaves and distributed among the clan members for feasting. "We gave one leg to one family, the other leg to another. One family got a hand or an arm with the ribs attached, another the back. But I kept the head because it belongs to the family that killed the *khakhua*. We cooked the flesh like we cook pig,

The khakhua *killers and eaters, brothers Bailom (left) and Kili-Kili, holding the skull of Bunop, the friend they killed and ate, believing him to be a* khakhua.

placing palm leaves over the wrapped meat together with burning hot river rocks to make steam."

The Korowai tie vines tightly around the arms and legs so that the meat swells up when cooked, making it easier to pull from the bones. Again, as I had asked Kornelius, I ask Bailom if human flesh tastes like pork. Bailom shakes his head. "Human flesh tastes like young cassowary," a local ostrich-like bird. At the cannibal feast, only the teeth, hair, bones, nails and penis are left uneaten. "I like the taste of all the body parts, but the brains are my favorite," Bailom says, prompting Kili-Kili to nod agreement, his first sign of life since he arrived at the hut.

Bailom feels no remorse. "Revenge is part of our culture, so when the *khakhua* eats a person, the people eat the *khakhua*. It's normal. I don't feel sad I killed Bunop, even though he was a friend." No mercy is shown even when the *khakhua* is a close relative. After capturing him clansmen bind him with rattan and drag him through the jungle for up to a day to a stream near the treehouse of a friendly clan. It is they who then kill and eat the *khakhua* in a grisly ritual exchange of human flesh between allies.

Bailom admits to killing four *khakhua*. And Kili-Kili? He laughs coldly and murmurs a reply. "He says he'll tell you now the names of eight *khakhua* he's killed," Bailom answers, "and if you come to his treehouse upriver, he'll tell you the names of the other twenty-two."

I ignore the ghoulish invitation. "When you've finished eating the *khakhua*, what happens to the bones?"

"We place them by tracks leading into the treehouse clearing, to warn our enemies. But the killer can keep the skull if he wants. After we eat the *khakhua*, we beat loudly on our treehouse walls all night with sticks to warn any of the *khakhua* who might be nearby and want revenge that we're a powerful clan."

As we walk back to the hut, I remind Kornelius that it was here that he ate human flesh. "It was Bailom who gave it to me," he says.

"If it had been the flesh of Bunop, your porter, and you knew, would you have eaten it." He shrugs and disappears into the darkness.

It takes me a long time to get to sleep, despite my fatigue, haunted by Kili-Kili's inhuman stare, the matter-of-fact description by Bailom of his *khakhua*-killing feats, and Kornelius's admission that he had eaten human flesh. I wonder what my reaction would be if a Korowai offered it to me. My throat clenches at the thought.

At night in Yafufla, Kili-Kili (left) and Bailom with Bunop's skull.

Kirk Huffman, an Anglo-American anthropologist, is a good friend and a world expert on Melanesian culture. He told me before I left on this journey that cannibalism had been common among many Melanesian cultures such as the Korowai up until the twentieth century when horrified colonial officials worked to stamp out the practice. "It was usually associated with magical forces, but there's nothing like the *khakhua* anywhere else," he said. "The Korowai probably keep their children away from the cannibal feast, not because they're squeamish, but because children are easy prey for evil spirits, and they want to protect them from any that might be hovering around when the adults are eating the *khakhua's* flesh."

The next morning I have to grapple with another horror. Kornelius arrives with a six-year-old boy named Wawa who is naked except for a necklace of beads. Unlike the other village children, boisterous and always smiling, Wawa is withdrawn and his eyes seem clouded by an emotional pain too deep for a child.

Kornelius wraps a comforting arm about the sad little boy. "When Wawa's mother died last November, I think she had TB, she was very sick, coughing and aching, people at his treehouse suspected that he was a *khakhua*.

Wawa, the little boy at Yafufla village whose parents died and who was then branded by villagers as a suspect khakhua *because they believed he might have killed them with sorcery. Wawa was treated as a pariah and is clearly terrified. When he grew older his relatives feared that villagers would kill and eat him as a* khakhua. *His relatives asked the author to rescue him and take him to Jayapura.*

His father had died a few months earlier, and they believed he used sorcery to kill them both. His family was not powerful enough to protect him at the treehouse, and so this January his uncle escaped with Wawa, bringing him here, where the family is stronger."

Despite the pain in Wawa's eyes, Kornelius does not think he truly grasps his plight. "I don't think he fully understands that people at his treehouse want to kill and eat him, though they'll probably wait until he's older, about fourteen or fifteen, before they try. But while he stays at Yafufla, he should be safe."

Kornelius escorts Wawa to the safety of his other uncle's hut, and then gives the command for us to leave. As the porters heft our equipment and head toward the jungle, he tells me: "We're taking the easy way, by pirogue." It is a large canoe hacked out of a tree trunk. He has asked Bailom to join the porters on the march. "He knows the clans upriver better than our men from Yaniruma."

Before he leaves, Bailom shows me his bow, a curve of hardwood tied with a length of springy rattan as its string. The arrows he carries are each a yard of slender bamboo lashed with vine to an arrowhead designed to kill a particular prey. The pig arrowheads are broad-bladed, those for birds long and narrow. The fish arrows have been fashioned into a prong, while the arrowheads meant to kill humans and *khakhua* are each a hand span of cassowary bone with six sharp barbs carved on each side. This ensures the arrowhead will cause terrible damage when cut away from the victim's flesh. Dark dried bloodstains are splattered on the cassowary arrowheads, possible proof that Bailom had used them to murder men.

As Bailom hurries to catch up with the porters, carrying his bow and arrows in one hand and with a pack strapped to his back, I ask Kornelius if he is comfortable that we have a dedicated cannibal accompanying us. "Most of the porters have probably eaten human flesh," he says flatly.

Kornelius leads me down to the Ndeiram Kabur where we board a long slender pirogue. I settle in the middle, the sides pressing against my body. I am surprised when Kili-Kili steps into the pirogue, carrying his bow and arrows, and sits in front of me. I turn back to Kornelius, and raise my eyebrows as I point to Kili-Kili. Kornelius gives me a helpless shrug, as if to say that he had not invited him to come with us, but I suspect you do not disagree with this fiercest of men without some risk to your well-being.

A pair of Korowai boatmen stand at the prow, gripping long paddles, with two more at the bow. They push off, steering the pirogue close by the riverbank where the water flow is slowest. We are going upriver, and the water out in the middle surges by, threatening to tip us over each time the boatmen need to maneuver the pirogue to the other side to get around a sandbar. Paddling upriver is tough going even for the lithe muscular boatmen, and, to buoy their spirits, they frequently break into song timed to the rhythmic slap of the paddles against the water, a yodeling chant that echoes along the Ndeiram Kabur.

In mid-afternoon, the pirogue turns off the river into what seems to be a floating jungle, the trees partly submerged in the murky stream. A few minutes later, the boatmen steer the pirogue to the edge of a swamp forest, where they tie it to a tree trunk. Kornelius leads me about twenty yards into the jungle to a nightmarish sight, a bleached skull framed by two thigh bones. "The victim's name was Gennavi, and Kili-Kili killed and helped eat him because his clan believed him to be a *khakhua*," Kornelius says. "The clan left the bones here as a warning to other clans."

The hairs rise on the back of my neck as I glance around and see Kili-Kili standing behind me. There is not a flicker of emotion on his face as he stares at his handiwork, not even pride that he had rid his people of the witch-man whose bones are sinking slowly into the mud. As for me, I cannot get used to the idea that among these cannibals the victims knew their killers. The act of cannibalism is gruesome enough, but there is an especial horror with the Korowai eating the flesh of men who grew up with them, hunted with them and laughed with them.

☠

Along the riverbank, high green curtains of trees woven with tangled streamers of vine shield the jungle. The sun beats down on my head like a hammer as a siren scream of cicadas shakes the air. The day passes in a blur, but I come alert when the sting goes out of the sun as it slips below the jungle rim. Night descends suddenly. "We've gone slower than I expected," Kornelius tells me. "We're still about an hour away from where we'll camp the night."

Eager to get to the camp, the paddlers bend their backs with vigour. Suddenly, from around the bend, a terrifying sound erupts, frenzied screaming and yelling. Moments later, through the gloom I see a throng of naked men on the riverbank brandishing bows and arrows at us. Kornelius calmly murmurs to the boatmen to stop paddling. "They're ordering us to come to their side of the river," he whispers to me. "It looks bad, but we can't escape. They'd quickly catch us if we tried."

My heart thuds as I peer at the shadowy tribesmen, their uproar banging at my ears now that we are so close. Our pirogue nudges the far bank of the river as Kornelius tries to reason with them, shouting across the water. "We do not want to hurt you, we come in peace," he calls.

"Do you have a gun?" I ask. He shakes his head. I fight the temptation to feel very afraid, but fail.

By now, the sky has darkened. I realize that if the arrows start flying, I would have good cover if I dived into the river. It is flowing very fast downstream, and would swiftly take me away from this danger. But Yaniruma is a few days downriver, and I would have to float all the way, risking being attacked by riverside clans and braving the crocodiles.

Kornelius continues to refuse to obey their command that we come to their side of the river. "It's too dangerous," he tells me. In front of me, Kili-

Kili sits in silence, but I see that he has gripped his bow and arrows and seems coiled to spring into action should we be attacked. For the first time I am happy that this famed *khakhua* killer is with us.

A pair of tribesmen slip into a pirogue and paddle toward us, and we shine our torches on them. As they come near, I see they carry bows and barbed arrows. "Keep calm," Kornelius says softly. "If we panic or make a false move, we'll be in serious trouble. Our boatmen know this clan, and they've just told me they'd kill us."

As their pirogue bumps against ours, one of the warriors growls that *laleo* are forbidden to enter their sacred river, and that my presence defiles it. Korowai are animists, believing that rocks, trees and even rivers are spirits. The warrior demands that we give the clan a pig for sacrifice to absolve the sacrilege. The cost of the pig is three hundred and fify thousand rupiah, or about $40. So, it is a shakedown. I count out the money, and a warrior takes it, glances at the bundle and grants us permission to pass. The tribesmen on the other side fade into the night as our boatmen paddle to safety upriver.

I cannot imagine what use money is to those men. Kornelius gives me a relieved smile. "Whenever they get any money, and that's rare, the clans use it to help pay bride prices for Korowai girls living closer to Yaniruma. They understand the dangers of incest, and so girls must marry into unrelated clans, while the boys remain at their clan's treehouse."

About one hour further along the river, we spot Bailom and the porters waiting on the far bank with worried faces. "We're staying here the night," says Kornelius as the boatmen steer the pirogue toward land. I scramble up the muddy slope, dragging myself over the slippery rise by grasping exposed tree roots. Kornelius asks Bailom, who knows the angry clan, how they knew we were coming. "As we passed near their treehouses, they stopped us when they saw we were carrying the packs and demanded to know who and where you were."

"Would they really have killed us if we hadn't paid the money for the pig?" I ask. Bailom nods. "They'd have let you pass tonight because they knew you'd have to return downriver. Then, they'd ambush you, some firing arrows from the riverbank and others attacking at close range in their pirogues. You could not have escaped."

I shrug off the incident, glad that we have surmounted another hazard. The porters spread all but one of the tarpaulin sheets over our precious supplies. Our shelter for the night is rough and ready, four poles set in a square about three yards on each side. The poles are topped by a single sheet of tarpaulin, leaving the sides open.

Soon after midnight, a torrential downpour slams into us without warning, drenching us in a few moments. The wind sends my teeth chattering, and cold and soaked to the skin, I sit disconsolately hugging my knees. Seeing me shivering, Boas pulls my body against his. I stiffen at the intrusion, but surrender as his body begins to warm me. Deeply fatigued, I ignore the rain and rest in his embrace, and he in mine. As I drift into oblivion I have the strangest thought. This is the first time I have ever slept with a cannibal.

CHAPTER 5

We leave at first light, still soaked from the downpour. At midday the pirogue reaches our destination, a riverbank close by the treehouse of a Korowai clan that Kornelius says is still living in the Stone Age and has never seen white people. He is a veteran of scores of journeys to the Korowai, but is nevertheless wary: this is foreign territory for him too. He grins on seeing that our porters have arrived and have built a rudimentary bamboo hut for us.

Kili-Kili leads the way up the slope, carrying his bow and arrows, and I am once more glad that he is with us. "I sent a Korowai friend here a few days ago to ask the clan to let us visit them," Kornelius says as I swig greedily from my water bottle. "Otherwise they'd have attacked us by now."

I ask why they have given permission for the first time for a *laleo* to enter their sacred land. "I think they're as curious to see you, the ghost-demon, as you are to see them."

At mid-afternoon, after a bracing sleep, Kornelius, Boas and I trek to the treehouse, called a *khaim*, walking for half an hour through the jungle close by the river and fording a murky waist-deep stream. I know we are near when a young cassowary, perhaps a family pet, prances past. A large pig, flushed from its hiding place in the grass, dashes into the thicket.

The jungle soon gives way to crops planted by the clan, mostly clumps of banana but also spindly sugar cane and knee-high rows of sweet potato. In startling contrast to their ferocity, Kornelius says the Korowai are keen horticulturalists with the men priding themselves on their green thumbs.

They honor themselves as being "lords of the garden." The evidence is this healthy growth of crops.

Kornelius looks tense and wary. He points ahead to a treehouse over-looking the river. It perches about eight yards above the ground on a decap-itated banyan tree with several supporting poles forming the framework, its floor a densely woven latticework of thick bamboo strips split down the middle and boughs. Van Enk had written in his book that when building a new treehouse, the Korowai stuffed the holes of the supporting poles with magical leaves and grass stalks, believing that this prevents *khakhua* and *laleo* from entering the treehouse by climbing up them. This leaves me wondering if the Korowai will allow me, clearly a ghost-demon, to enter their home.

"The treehouse belongs to the Letin clan," Boas says. In contrast to Kornelius, he seems to be enjoying this adventure, leading a *laleo* to the treehouse of a clan he knew of but had never dared visit. (Korowai form into a widespread patchwork of what anthropologists call patriclans, clans inhabiting ancestral lands and tracing ownership and genealogy through the male line.) Kornelius points to the treehouse. "They're waiting for us."

I hear many voices murmuring above as I climb a slanted pole with foot notches called a *yafin*. Inside the *khaim*, wreathed in a haze of smoke split by beams of golden light, about fifteen naked youths crowd the front of the treehouse. As I enter, they peer at me without embarrassment or fear. Smoke from the hearth fires has seeped into the bark walls and sago-leaf ceiling, giving the treehouse a sooty odour. A pair of stone axes, several bows and arrows and a handful of net bags are tucked into the leafy rafters. The floor creaks as I settle cross-legged onto it.

The nooks and crannies are stuffed with the haunting bones from clan feasts, with spiky fish skeletons, blockbuster pigs' jaws, and the skulls of flying foxes and rats. They even dangle from hooks strung along the ceiling, near bundles of many-colored parrot and cassowary feathers. The Korowai believe that visitors, on seeing the bones and feathers, feel welcome, while recognizing that they have come to the home of a prosperous clan. There is nothing modern, no mirrors, radio, chairs, tables, cupboards, steel utensils, matches, books, TV, dishwasher, taps, phone or even lamps.

Four women and two children sit at the rear of the treehouse, the women refusing to look at me as they studiously weave tote bags from vine. "Men and women must always stay at different ends of the treehouse, and

A warrior climbing up to his treehouse. The Korowai live in treehouses because when there is a fight with another clan in the jungle clearing the young, old and women can quickly escape. It's also breezier that high up.

they have their own hearths," says Kornelius. "They each cook their own food." Each hearth is made ingeniously from strips of clay-coated rattan suspended over a hole cut into the floor. If a fire starts to burn out of control, it can be quickly hacked loose, to fall to the ground.

Straddling the gender dividing line sits a middle-aged man with a hard-muscled body, a bulldog face and a pair of holes bored into his nose to hold bat-wing struts, decorations worn during feasts. As Kornelius talks through Boas to the warriors, he grips his bow and arrows settled by his side and avoids my gaze, staring the other way, as if I were an illusion, soon enough to drift away like the smoke that curls up from the fire.

"A clan of six men, four women, three boys and two girls lives here," says Kornelius. "The others have come from nearby treehouses to see their first *laleo.*"

He is an old Korowai hand, and knows to keep his talk deliberately mundane in this first encounter, exchanging information about crops, the weather and past feasts. Now and then I catch the older man glancing at me when he thinks I am not looking, but he turns away when he sees me glance back. "That's Lepeadon, the clan's fierce man," says Kornelius. Although the Korowai do not have chiefs, the fierce men, called *khen-mennga-abul*, are born leaders, bristling with the ferocious spirit needed to lead their clan brothers on successful raids, or in clan fights over pigs or women. He certainly looks the part.

Throughout Melanesia the cult of the Big Man rules, even in vastly different cultures separated by thousands of miles of ocean. Powerful and charismatic men hold sway over their clansmen, not only in battle but also in the everyday minutiae of their lives. They are like those massive elephant seal bulls that dominate the beaches and form harems with many females. When the other bulls challenge their dominance, they risk a bloody fight, sometimes to the death.

In the New Guinea highlands, these Big Men rule with a heavy hand. Nevertheless, they are revered, with the names of the greatest resounding through the generations—as I had seen at Wamena, the three-hundred-

Lepeadon, the fierce man of the Letin people, on the balcony of his treehouse.

year-old mummy of the Dani's most famous Big Man still kept in the place of honor in the men's ceremonial hut.

This independent streak of the Korowai attracts me because we Australians like to think of ourselves as rugged individualists, our resistance to and distrust of the Big Men among us tempered in the early days of our nation when most of the settlers, including my maternal great-great-grandfather, were convicts hauled out from Britain in chains.

"Each man does what he wants in his own treehouse," Boas had told me earlier, "but when we fight other clans we follow the *khen-mennga-abul*, the fiercest man, into battle. My father is a *khen-mennga-abul*, and is feared and also honored by our people, just like Kili-Kili. But my spirit is not as powerful as his, and when we have to fight I always follow my father. When you meet him you'll see what I mean."

After listening to an hour of Kornelius's small talk, the fierce man moves closer and stares at me with a look that mingles suspicion with a hint of acceptance. Still unsmiling, Lepeadon says: "I knew you were coming and expected to see a ghost, our people were very afraid, but now I see you're just like us, a human."

As if to prove this, a youth suddenly lunges at me, trying to drag down my pants at the front. He almost succeeds, prompting much laughter from the other men. I join in, a little nervously, but keep a tight grip on my modesty. Reverend Veldhuizen had told me that the Korowai believed he was truly a ghost-demon, and it was only when they spied him bathing naked in a stream by a treehouse and saw that he came equipped with all the necessary parts that they realized he was a *yanop*, a human being. His shirt and pants baffled the naked Korowai who were unable to comprehend his clothing. They called it *laleo-khal,* or ghost-demon skin, a magical epidermis that Veldhuizen could remove or put back on at will.

"We shouldn't push this first meeting too long," Kornelius tells me as he rises to leave. Lepeadon follows us down the ladder to the ground, and grabs both my hands. He begins bouncing up and down and chanting, "*Namaiyokh,*" or friend, at each thump of his feet on the grass. I follow him in what seems a ritual farewell, and he increases the pace until it is frenzied before he suddenly stops, leaving me breathless.

"I've never seen that before," Kornelius admits. "We've just experienced something very special." He's right. In four decades of journeying among remote tribes, this is the first time I have been with a clan that has never before seen a white person, and my eyes become teary as we return to our hut. I can thank Kornelius's careful preparation for the first contact being peaceful, and not a potentially deadly confrontation like the river ambush.

At camp, Boas takes my hands and begins the same chant as Lepeadon. "*Namaiyokh, Namaiyokh, Namaiyokh,*" he repeats, though he leaves out the bouncing up and down. "It's a ceremony we always perform when a friend or a relative comes from far away, and whom we've not seen for a long time," Boas tells me.

"So, Lepeadon in that one meeting went from thinking I'd be a ghost to realizing that I was a person, just like him but with a different colored skin and features. Maybe that prompted him to claim me as one of his own."

Boas smiles. "I think he was relieved that you weren't a ghost. He took a big risk in agreeing for you to come to his treehouse, because you might have got violent and destroyed it and his people. He was so happy that even though you weren't one of his relatives, or a clansman, he treated you as if you were."

☠

Dinner is a plate of rubbery noodles mixed with canned beans suffused with a metallic taste. Boas scorns our food, preferring to eat with his hands a pile of the Korowai staple food, sago, and a big spider he grills on a river stone placed in the burning embers of the porters' fire. When he offers to share it with me, I spurn the spider but try the sago. It tastes like burnt flour, and is so dry that the lump sticks in my throat. Boas laughs as I half-choke on it, and then free the sago with a glass of water causing it to slip down my throat. "Sometimes I wonder if you're truly a *yanop*, a human, because you eat food I'd only give to our pigs, and even they might not eat it," he says with a grin. A hint of a smile hovers around Kornelius's eyes, but he says nothing.

Next morning, four young women arrive at our hut just after seven o'clock carrying a squawking green frog, several chirping locusts, a lizard and a spider they had just caught in the jungle. "*Manotropo laleo,*" a pert teenager says to me. It means, "Good morning ghost-demon."

"They've brought your breakfast," Boas smiles, enjoying his joke. Two years living in a Papuan town, far from his treehouse, must have taught him that we *laleo* wrinkle our noses at Korowai delicacies. The teenager, whose breasts have not been deflated by motherhood like those of the married

Korowai women, mimics the act of eating the live locust, her eyes shining with pleasure as she pretends to pop it into her mouth. She then offers it to me, and seems bemused when I shake my head, politely declining the offer. Boas is amused by my reluctance to try jungle food, and when the girl hands the locust to him, chews it with exaggerated pleasure to show me what a fool I was to spurn the treat.

The girls must have left the treehouse soon after dawn because they have already collected firewood carried in hand-woven vine bags strung around their foreheads. Life is tough in the jungle, and all are wiry with whipcord muscles. In contrast to the naked men, Korowai girls wear short grass skirts made from strips of dried palm frond that perch on the hips covering the genitals and buttocks, perhaps to keep the men's lust in check or dampen sexual jealousy. I have seen this difference elsewhere many times among remote people.

The only tribe I have encountered where the men and women both go entirely naked was the Korubo in a remote Amazon jungle, but I know of Sudanese tribes and Australian Aborigines where both sexes once wore no clothes. I never once saw the Korubo men sexually aroused, even though I spent a week with them, and suspect that it stems from the same effect experienced at nudist colonies where the constant sight of naked bodies commonly deflates rather than excites sexual desire.

One of the girls holds at her waist a lean muscled hunting dog. It has sharp black eyes and is covered with patches of brown and saffron hair, and resembles a small basenji, the dog used by the ancient Egyptians to stalk game. It looks much like the hunting dogs I saw accompanying pygmies on a hunt in a Congo jungle. The prettiest girl holds against her jutting breast a small pig, a family pet until it grows to full size when it will be eaten, or more likely used as part of a bride price.

"The Korowai rarely eat their pigs, using them as barter, as proof of their wealth, as compensation in a dispute or as a bride price," Kornelius tells me as I pat the piglet, prompting a lovely smile from its guardian. "A highly desirable girl can cost the bridegroom's family at least ten pigs and a string of dogs' teeth up to twenty yards long."

The girls have circles of scar tissue the size of large coins running the length of their arms, around the stomach and across their breasts. "The marks make them more beautiful," Boas says.

He shows me how they are made, placing coils of coin-size bark on the

Korowai women emerge from the jungle with breakfast including a frog and cicadas.

skin and setting fire to them. It seems an odd way to add beauty to the female form, but no more bizarre than tattoos, the torture of stiletto-heel shoes, cosmetic surgery or the not-so-ancient Chinese custom of slowly crushing the feet bones of infant girls to make their feet as permanently small and thus as erotic as possible.

"What do you look for in a girl you'd like to marry?" I ask Boas.

"She must be hard-working, have a strong body, know how to get plenty of food from the jungle and know how to raise children," he says. Do her looks matter? "Not much, though if she's all that and is also pretty then she'll bring to her father a good bride price."

What does a girl look for in a husband? "He has to be a good hunter so that she and their children will never go hungry. That's all."

"Are you married?"

"Not yet. I haven't found a girl I like that much. But I'll have to soon because my father wants to see my children before he dies."

☠

There had been a downpour overnight, leaving the forest rain-speckled and exuding a giddy odor of dank earth and foliage. With the morning mist drifting between the trees, Kornelius, Boas and I once more trek through the jungle to the treehouse where Lepeadon is waiting for us. In the clearing the sun has yet to reach the high tree-line, and the sky is raw grey, luminescent along the treetops.

The Letin clan's other members have long gone on their daily search for food, the women and children collecting edible roots, plants, insects and lizards and the men seeking wild pig, snakes, cassowary and birds. But Kornelius had earlier sent Boas to the *khaim* to ask Lepeadon if he would stay and talk to me.

I want to speak with the clan's fierce man about the Korowai belief in the supernatural and their conception of cosmology. Kornelius and Boas had told me some: the Korowai believe the universe to be made up of three concentric circles. In the inner circle reside the people, animals, *laleo* and *khakhua*. The second circle holds the land of the dead who live within the same clan system as the living. As in life, each clan's territory is clearly defined. A broad pathway cutting through the jungle connects the two circles, allowing those who have just died to walk across the divide and join their deceased relatives.

Lepeadon looks puzzled when I ask if there is a separate place for people who have acted badly in this life where they can be eternally punished. "He doesn't understand what you mean, because all Korowai, good and bad, and even the *khakhua* go to the second circle when they die," Kornelius explains.

Surrounding these two circles of the living and the dead is the third circle, a never-ending stretch of great water populated by huge fish, Lepeadon explains. "We call it 'the endless great water.'"

Although the Korowai are only about one hundred miles inland from the Arafura Sea, which has its share of enormous sharks and massive salt-water crocodiles, between the treehouse people and the ocean reside fierce headhunter tribes such as the Citak and Asmat. It is unlikely then that any Korowai in centuries past ever broke through such a murderous gauntlet to reach the sea and returned home to tell of its wonders and horrors. It is more likely that over a very long time, word was passed from tribe to tribe of the great and dangerous expanse of water to the south, and that the Korowai wove this startling knowledge into their view of the universe.

"I've been to the third circle where giant fish swim in a great stretch of water that takes many days to cross in a canoe," I tell Lepeadon, attempting to impress the clan's fierce man. "Do you want to go there?"

The war chief trembles. "Never! I don't want to go beyond the land of the dead. When I die my soul will escape through my skull and then it will travel through the jungle across to the land of the dead. There I can hunt, marry, eat pig and cassowary, but I can never have children there."

"Are there *khakhua* in the land of the dead?" I ask.

"Yes, but they never bother anyone there."

Again, I try to impress him. "I can take you to the third circle."

He smiles and nods, but doesn't follow up the offer, implying that there is no need for such a journey because no Korowai that he knows has ever questioned the truth of the revelation. So, there is no need for me to offer him proof or for him to seek it.

This rigid belief in their creation myths seems common among Stone Age tribes, because whenever I have journeyed among them, I have found no evidence that the clans question the divine revelations of their shamans. Their rock-hard faith contrasts with the continual ferment of doctrinal challenge among followers of the three great religions formed over the centuries in the deserts of the Middle East—Judaism, Christianity and Islam—with the espousal of countless heresies and the formation of countless breakaway sects.

I have experienced a similar lack of interest in modern culture among other Stone Age tribes even though the way I look and act make it obvious that there is a vast difference between the way I live and the way they live. During my stay with the Letin clan, neither Lepeadon nor any other person asks me a single question about my way of life. Perhaps, intuitively, they know that the leap across the culture gap that separates us is too vast for them even to attempt. Or perhaps they just do not care.

Three years earlier I had visited the Korubo, a very violent Stone Age tribe. They inhabit an Amazon jungle closed to almost all outsiders by the Brazilian government which seeks to protect their unique way of life. I was given the rare privilege of being with the Korubo by their protector, Sydney Possuelo, director of the department that oversees remote Indians. On our last day with the clan he tried to explain cars to them in a way he thought they might understand, saying that they were like small huts that had legs and ran very fast. I watched the Korubo shake their heads and look at him with disbelief at what sounded to them like the babble of a madman.

Although Lepeadon shows no interest at all in my spiritual beliefs, he is willing to explain his beliefs, though at first he lets the details dribble out.

I refer repeatedly to van Enk's book to show him that I know something of Korowai spirituality, and this encourages him to open up. He tells me that above the circles of the Korowai living and the dead hovers the sky with its sun, moon, stars and planets, but implies that these have no connection with the Korowai's supernatural or spiritual beliefs.

I ask him about a Korowai belief that dead people can return to the land of the living if a tribal council of the deceased in the second circle gives them permission, usually to be reborn as a baby. The body of the dead person is then given a ceremonial burial in the land of the dead. He nods agreement, but cannot or will not explain to me why a dead person is allowed to be reborn. He confirms that a dead person's shadow hovers about for a very long time in the land of the living, and that there are Korowai men who possess special powers, able to change into animals and then return to their original bodies.

"Do you have such a powerful man in this treehouse, or nearby?" I ask.

Lepeadon stares deeply into my eyes. "It's too soon for you to know such a thing. Come back again another time and I will tell you."

Kornelius touches my arm. "Our porters say Lepeadon is such a man."

"Then, one day we'll have to come back."

"If you want then you should. I sense that Lepeadon likes you."

I tell Lepeadon that most of my people believe in the one God, a credo he finds hard to understand. "There are small gods in everything around us, the trees, rocks, water, animals," he tells me. But, he adds, the Korowai do have a supreme spirit named Ginol Silamtena who created the present world, his fifth effort, having destroyed the previous four.

Around the fires for as long as the collective tribal memory reaches back, old people have impressed upon the younger ones that white-skinned ghost-demons would one day invade Korowai land. Once the *laleo* arrive, Ginol warned, he will obliterate this fifth world. The land will split apart, there will be fire and thunder, and mountains will drop from the sky. This world will shatter and a new one will take its place.

The prophecy's accuracy saddens me, and I admire Lepeadon all the more for allowing me, one of the feared *laleo*, to come into his universe when his beliefs warn that my presence will surely bring disaster crashing down on the heads of his people.

Van Enk spent several years among the Korowai learning their language and studying their religion, and yet he admits that he has only an elementary knowledge of their spiritual beliefs. I am not to be given any more. Lepeadon's wary smile hardens into a glare as he tires of my questioning. Without another word, the fierce man picks up his bow and arrows, climbs down the ladder and disappears into the jungle. Kornelius responds with a shrug.

At midday we return to our hut and there Kornelius decides that we will not return to the treehouse until the next morning. The Korowai often remain in the jungle until it is nearly dark, hunting and gathering food, and once home they rarely venture out of their treehouses at night because that is when they believe that the evil spirits lurking in the jungle are most active. "You risk having the evil spirits kill you with magic if you leave the treehouse while it's dark," Boas tells me.

With the night comes a silent black void that swarms about our hut like a malevolent spirit. Beyond the feeble light of our torches and the smouldering fireplace, dozens of male fireflies invade the clearing, dancing through the darkened air like wicked fairies, pinpricks of pulsating silver light tracing their zigzagging path.

I smile at their spectacular seduction display, but Boas and our Korowai porters huddle around the fire, their gaze turned on the flames, because to them and to the treehouse dwellers the forest at night is a terrifying place, and even the sight of it shakes their souls. And why not? Their fear is much like that of our ancestors who lived in caves thousands of years ago. They too crouched over comforting fires in their nighttime sanctuaries, dreading the demons, hobgoblins, witches, banshees and trolls prowling in the blackness beyond the caves.

CHAPTER 6

The Korowai divide the day into seven distinct periods: dawn, sunrise, mid-morning, noon, mid-afternoon, dusk and the night. They do not count the days, weeks or months, but allow each day to pile on the one before it so that none, not even Boas, know how old they are.

They use their bodies to count numbers. The next morning at the *khaim*, when I ask Lepeadon to show me how, he ticks off the fingers and thumb of his left hand starting with the smallest, and continues to count as he touches his wrist, forearm, shoulder, neck and forehead. He then moves down the other arm until he reaches the little finger on the right hand. The tally comes to nineteen. Anything greater than that the Korowai refer to as "many."

At mid-morning, I go with the clan to the sago palm fields to harvest their staple food. Because the sago palm grows slowly, each clan has many fields under cultivation. Sometimes the fields can be an hour or more away, but this time I am lucky because the clan takes me to a field just ten minutes walk from the treehouse. The warriors carry bows and arrows, as they always do when they step even one foot beyond the treehouse, to shoot game or fight off enemies. The women have vine bags strung about their foreheads to collect any edible plants they spot along the way.

I follow Lepeadon and the others to a muddy stream that meanders through a scattering of palm trees shadowed by high rainforest giants at the foot of a steep hill. Lepeadon and a younger man, both with the taut muscular bodies of athletes, select a stout sago palm tree, about twelve

At Letin, the Stone Age Korowai village. The men always carry their bows and arrows into the jungle. They also carry stone axes to chop down palm trees.

feet high, which has taken at least ten years to mature and is just about to flower, ensuring that the amount of starch inside the tree is at maximum level. They chop it furiously, each with an ax that defines an entire epoch of human existence, the Stone Age. The ax is a fist-sized chunk of hard dark stone traded from clan to clan for more than a hundred miles from the mountains, sharpened at one end and lashed with vine to a short slim wooden handle. The stone is so scarce that the clan has just the two stone axes.

It takes more than an hour to chop down the palm tree, the men working in shifts, two by two, for about fifteen minutes at a time. As it topples with a thump, the warriors let out a blood-curdling scream and, led by Lepeadon, charge the fallen palm. They fire their arrows into it as if they were skewering a giant enemy. A snake had leaped from the palm tree as it fell and is killed. They then perform a war dance about the palm, whooping and yelling in triumph. Boas shrugs when I ask him the meaning of this ritual. "That's what we've always done, but I don't know why."

The men take it in turns using the stone axes to pummel the sago pith to a stringy pulp which the women sluice with water through a filter of bark

Young warriors at Letin doing a war dance.

to produce a dough they fashion into large balls and wrap in sago leaves. At the treehouse, they mould the sago into bite-size pieces and grill it in the hearth.

The Korowai know only one way to make fire, and I expect to see them rub two sticks together, the legendary way for Stone Age people. But Lepeadon uses a far more efficient method, looping a length of rattan about a stick that he is standing on and rapidly pulling it to and fro. The friction swiftly produces tiny sparks that set fire to a bundle of shavings placed by the stick. Blowing hard to fuel the growing flame, he then places the snake under a pile of burning wood.

Korowai always share any food—to refuse to do so is a clear sign of enmity—and Lepeadon offers me a chunk of the charred meat. It tastes like tender chicken.

On our return to the treehouse, each time we pass a big banyan tree, the warriors with great excitement slam their heels against the knee-high struts circling the tree's base, producing a hollow thumping sound that travels far across the jungle. "That lets the people at the treehouse know they're coming home, and how far away they are," Kornelius says.

The days with the Letin clan pass swiftly. When I feel they trust me, I ask when they last killed a *khakhua*. The fierce man uses as an indicator the most recent sago palm feast, when up to a thousand Korowai at a time, a quarter of the population, gather to perform traditional dances, eat vast quantities of sago-palm maggots, trade goods, chant fertility songs and let the marriage-age youngsters eye each other. According to our porters, that dates the killing and eating of the witch-man here to just over a year ago.

"My uncle was sick, his body ached all over and he knew he was dying," Lepeadon tells me. "When it was close to the end, he told me the name of a person from a treehouse near here, the *khakhua* who was killing him. I knew him well and was sad that a *khakhua* now was inside him and killing my uncle. I felt great hatred for that *khakhua* and longed to kill and eat him, to avenge my uncle. The day after my uncle died we buried him beneath the treehouse. I asked the *khakhua* to come on a hunt with us the next day, and when we were in the jungle we grabbed him and tied him up. He struggled, and I hit him on the head with an ax, not too hard. I didn't want to kill him, just quieten him. We took him to the river, and fired our arrows into him. Then, we cut him up, wrapped the body parts in leaves and brought them back to our treehouse for the feast."

"Where did you cook him?"

Lepeadon points to a clearing by the river, about fifty yards from the treehouse. "We cooked him the same way we cook pigs."

"Did the meat taste like pig?"

"No, it tasted like *khakhua*."

"Did it taste like young cassowary?"

Lepeadon pauses for a few moments. "Similar, but not exactly the same. It's less stringy and there's more blood."

"Do you enjoy the taste?" I ask the war chief, much as if I were discussing the merits of the meat we were preparing at a barbecue in my Sydney backyard.

"Yes, all the people like to eat *khakhua*. Most people like the brains the best, but I like the tongue."

"Do you eat all the meat at the same meal?"

"No, only some of it. The head, a leg and the back. The rest we give to a couple of clans who live near us. When they kill a *khakhua* they give us a share."

"Does the man's family accept that he was a *khakhua*?"

"I've never asked them," Lepeadon shrugs.

"Is it likely they'll target one of your clan as a *khakhua* and kill and eat him in revenge?"

"If one of us was a *khakhua* then they would, and I'd help them."

"What if you were the *khakhua*?"

"If that were true then I would already be dead because the *khakhua* would have secretly killed me and taken over my body."

Kornelius tells Lepeadon that we must leave the next morning. He asks me to stay longer, but I tell him with regret that it is not possible as I must return to Yaniruma to meet the Twin Otter on a fixed date. He looks beyond me, out over the jungle, hopefully accepting in good faith the need for my departure, but surely not understanding the way we *laleo* must jump to the command of clocks and calendars.

Just before we depart he grabs my arm. "Come back again, *laleo*," he says with a rare smile.

At sunrise the next morning, as I climb aboard the pirogue at the river's edge by the treehouse, Lepeadon squats at the top of the sloping bank, but refuses to look at me. When the boatmen use their paddles to push away into the river, he leaps up, scowls, thrusts a cassowary-bone arrow across his bow, yanks on the rattan string and aims the barbed arrow at me. For a

Letin clan in one of their treehouses. Lepeadon, the fierce man, is at the extreme left.

few moments he lets me fear that he is going to skewer me, and then smiles and lowers the bow, a fierce man's way of saying goodbye.

In mid-afternoon, the pirogue turns off the river and heads up a murky stream with the rainforest canopy arching overhead. About ten minutes later, the boatmen tie the pirogue with vine to a tree by the sloping bank. Boas leaps out and disappears, eager to see his family. Kornelius and I follow, far more slowly, and after an hour's trek we reach a clearing about the size of two football fields and planted with banana trees. Dominating it is a treehouse that soars about twenty five yards into the sky. Its springy bamboo floor rests on several natural columns, tall trees cut off at the point where the branches once flared out and took leaf.

Boas is waiting with his father, Khanduop, a middle-aged man clad in rattan strips about his waist and a leaf wrapped about his penis tip. He grabs my hand, thanking me for bringing his son home. Lepeadon looked like the fierce man he was, but Khanduop at first glance has a clerkish mien, a small slight man with the hint of a grin and few visible muscles. Looking closer, I see a fire burning deep in his eyes. I wonder whether he might be similar to what the ancient Vikings called a *berserker*, a man whose ferocious spirit and willingness to charge into battle without a thought for his own safety cowed the warriors who fought him.

To celebrate the return of the prodigal son, Khanduop decides to kill one of the family pigs, a large dark porker that squeals in terror as it is dragged to the edge of the jungle. His wife watches with tears streaming down her cheeks. Boas, her stepson, tells me, "She loved the pig, and treated it as if it were one of her children," even though she knew that one day it would be killed and eaten, or traded away to another clan.

Khanduop draws back his bow and shoots a single broad-bladed arrow into the pig just below the ribs. After a final heart-wrenching squeal it dies. Then Bailom, with superhuman strength, carries it piggyback up the steep notched pole and into the treehouse. Inside, like the Letin treehouse, many bones of animals eaten at past feasts have been pushed into the crevasses and folds of the bark walls.

Using a stone ax, Khanduop expertly carves up the pig, much as he must have carved up the bodies of the many *khakhua* he has killed and eaten. Some of the bloodied parts are placed by the fireplace to be consumed by the clan, while Khanduop's wife wraps others in banana leaves to be sent to neighboring clans.

Yakor, a tall kindly-eyed warrior from a treehouse upriver, squats at the fire with Khanduop, Bailom and Kili-Kili. Boas's mother died several years ago, and Khanduop is now married to Yakor's sister. As they consume the cooked pig, the talk turns to cannibal feasts they have enjoyed. Khanduop's eyes light up and he smiles widely when he describes feasting on human flesh. He has dined on many *khakhua*. "The taste is the most delicious of any meat I've eaten," he says.

I leave the men to their meal and step out onto the narrow veranda, high above the clearing. From here I can see why the Korowai build these towering leafy fortresses. Their principal use is to thwart invaders, but there are many other advantages to high-rise jungle living. Treehouse inhabitants are safe from the evil spirits that flit about the forest floor at night, and escape the malarial mosquitoes that forever seek human blood below. The air up here is fresher, cooler, breezier. From the veranda the men can shoot birds

Boas (left), in his yellow bonnet, and his father, Khanduop, at the latter's treehouse just before Boas left for Yaniruma, a settlement downriver.

flying by, and also gain pleasure by contemplating their beloved gardens in the clearing below.

The Twin Otter is due at Yaniruma the following afternoon, and so the next morning the porters depart for the river carrying our remaining supplies. Before I leave, Khanduop asks to talk about the Korowai future. "Boas has told me he'll live in Yaniruma with his brother, coming home just for visits," he sighs. His gaze clouds with grief. "The time of the true Korowai is coming to an end, and that makes me very sad."

Bailom is returning upriver to his treehouse, accompanied by Kili-Kili. "Come back one day to my treehouse. We'll hunt together, and eat frogs and spiders and pigs."

As Bailom and Kili-Kili head toward the jungle rim and are quickly swallowed up by the trees, Boas gives his father a farewell smile and walks with me back to the pirogue for the two-hour journey downriver to Yaniruma. He has put back on his yellow bonnet, as if it were a visa for his reentry into the twenty-first century.

☠

When I visited that other violent Stone Age clan, the Korubo, in the Amazon with Sydney Possuelo, he told me that he had been troubled for decades by this question of what to do with the world's few remaining Stone Age tribes, whether to yank them into the twenty-first century, or whether to leave them untouched in their jungles with all the barbarity of their traditional lives. "I believe we should let them live in their own special worlds," he told me, "because once they go downriver to the settlements and see what to them are the wonders and magic of our lives, they never go back to live in a traditional way."

So it is with the Korowai. They have at most one generation left in the Stone Age, their customs violent and savage to us, but, to them, the way humans should and must live with honor. Year by year, the young men and women will drift to Yaniruma, Manggel and other settlements downriver on the Ndeiram Kabur until there are only aging clan members left in the treehouses. That is when their god Ginol's prophecy will reach its apocalyptic fulfilment, and thunder and earthquakes of a kind will destroy the Korowai world forever.

At Yaniruma, Boas and I hug farewell. "I like you, ghost-demon," he smiles as he hands me the stone ax he has been carrying since we left his father's treehouse. "Take it home so you'll always remember the Korowai."

As the plane shoots into the air, I peer at the Ndeiram Kabur below, twisting like a snake through the dense jungle. From up here it looks impassable. A few minutes later, I spy Khanduop's treehouse, and imagine that the ant-sized people scurrying about in the clearing are waving at me.

I settle back in the canvas seat, eyes closed, wondering why I already miss my cannibal friends so much—Boas, Lepeadon, Khanduop, Bailom and even the dreaded Kili-Kili. I had gone into the Korowai jungles expecting to find bug-eyed bloodthirsty monsters thrust beyond the pale by willingly breaking the most sacred taboo, eating human flesh. Instead, I found a people living as best they could in the way they were raised, even though they were mired in a tradition of perpetual violence. Korowai parents are anxious about the health and future of their children. Boys and girls fall in love. Men and women trudge off each morning for a full day's work, and then return home at night to a tasty meal and a chat around the fireplace before going to bed. Just like most of us. I realize that Kornelius was right. Had I been raised as a Korowai, then I too would have been a cannibal.

Saintly Cannibals of the Ganges

CHAPTER 7

I was not surprised by the Korowai using the supernatural to rationalize their deep need to understand and explain mysterious deaths probably caused by bacterial disease. It was just another step in their minds for them to then execute and eat *khakhua* to combat these dark forces from the otherworld. But the impetus for my next journey among the man-eaters was so implausible that I listened in disbelief when told by a friend, a religious scholar in my hometown Sydney, about a unique form of cannibalism. In India, in the holy city of Benares, he said, I would find a Hindu sect named Aghor, whose sadhus, or holy men, were saintly cannibals. They eat human flesh as the supreme demonstration of their sanctity.

This seemed an improbable, perhaps impossible, union of the most extreme of extremes, and so I sought the advice of Ron Barrett, an assistant professor in the Department of Anthropological Sciences at Stanford University in California. He is a devotee in the sway of an Aghor saint who has an ashram not far from San Francisco. I phoned Ron at Stanford. "Little is known of the Aghoris outside India, but they're well known there," he told me. "Even though their holy men have a reputation for doing wild things like eating human flesh, and drinking a lot of alcohol, and living at cremation ghats, a former Prime Minister, three state governors and many other notable people are among our members. But there are valid religious reasons for these practices. By rejecting the norms of their society, and practicing what that society finds abhorrent and forbids, in effect that allows the Aghor holy men to move outside their own society and embrace all humanity."

I also found a book—*Aghor: At the Left Hand of God*—published in 1986 by another American devotee of the sect, Robert Svoboda. The cover showed a ghoulish but typical representation of Kali, the revered Hindu mother-goddess, wife of the paramount god Shiva the Destroyer. She wore a necklace of bloodied human heads, stood triumphantly on a slain enemy, poked out a blood-splattered tongue and held a skull in one hand, a killer's noose in another, a sword in her third hand, and a pair of clippers in the fourth. Kali seemed a worthy choice to be the patron deity of a sect whose holy men can be cannibals.

As Svoboda relates, in Hindu mythology, when Kali died, a grief-stricken and enraged Shiva perched her body on his shoulders and went on a wild whirling dance about the world, threatening to destroy it. Another power-ful god, Vishnu, threw a knife at Kali's corpse to stop Shiva's maddened dance, and the body broke into more than fifty pieces, which were flung about the planet. Her little toe fell by a river in Bengal, and was discovered much later, prompting her devotees to build a temple and a village there named Kalikata, or Calcutta. Ironically, that temple, where goats are still sacrificed daily to this bloodthirsty deity, stands next to the original sanctu-ary set up by Mother Teresa for men and women abandoned in the streets as they lay dying.

On the book's back cover is a painting of an Aghor saint, naked and holding a human skull. Scattered about him are human skulls and bones. The author, Svoboda, defined Aghor/Aghori as "the most extreme of all the Indian sects, concentrating on forcible conversion of a limited human personality into a divine personality."

That is an intriguing way to explain a sect that reveres holy men who eat human flesh. *Aghor* means in Hindi "that which is not difficult or ter-rible," which seems to me the exact opposite of the sect's surmounting philosophy.

Svoboda without embarrassment recounts the cannibalistic behavior of his guru Aghori Vimalananda who claimed that "an Aghori is beyond the bound of the earthly shackles." To prove this, Vimalananda revealed to Svo-boda that he and his son had eaten human flesh: "I used to wait at the fu-neral pyre until the skull would burst—it bursts with a fine pop—and then I would rapidly, to avoid burning my fingers, pull out parts of the brain, which would be a gooey mess, partially roasted by then, and would eat it."

This reminded me of Bailom's revelation that he favored the brain as the tastiest part of the human body, and this similarity to the Aghori guru's choice gave validity to the possibility that at least some of the sect's holiest priests were cannibals. Until this revelation, very little had surprised me about India during my many visits over the decades to the subcontinent including its recent astonishingly swift transformation from the medieval age to the computer age, as if a new-age Indian had suddenly sprung from the chrysalis. India is rapidly emerging as one of the twenty-first century's giants, basing its charge toward greatness on its mastery of high technology, graduating tens of thousands of scientists from its universities each year. Its economy is sizzling, with the latest annual growth rate nudging 10 percent.

The Western media has applauded this transformation, but I knew it was a chimera. Just one million IT workers bring in more export income than several hundred million Indian farmers. One uses a laptop humming with invisible connections to the outside world, the other a plow turning over ancient land, and yet they have far more similarities than differences. Most are devout believers in the millennia-old superstitions that still hold one billion Indians, rich and poor, literate and illiterate, in their sway.

It seemed certain that if there were cannibal holy men in India then I would find them, as my friend suggested, in Benares, one of the world's oldest continually inhabited cities, its origins dating back more than three thousand years. On a visit at the end of the nineteenth century, an impressed Mark Twain wrote that, "Benares is older than history, older than tradition, older even than legend, and looks twice as old as all of them put together."

It is the central repository of the country's ancient cultures in all their myriad and bizarre forms. "Benares is the heart and soul of India," a friend, Navreet Raman, UNESCO's representative there, once told me.

So, as Hindu pilgrims have done for more than 2,500 years, I journey to ancient Benares, India's holiest city. Hindu creed preaches that anyone who is cremated in Benares and whose ashes are then cast into the Ganges River, which flows in a crescent around the city, will immediately enter into *moksha*, or enlightenment. This is a kind of fast dissolve of the soul into the eternal ether, ending forever *samsara*, the relentless cycle of birth, death and rebirth.

☠

On my pilgrimage to the Hindu holy city, I enter India through Bombay or Mumbai. Like India itself, Bombay is a place of startling extremes, mansions of the wealthy lining potholed streets, looming over dirt-poor families who sleep in the streets, huddled along the grubby pavements. There, they eat, sleep, laugh, fight and even beget a new generation, sometimes in full view of passers-by. On the way to the city center I also pass the more fortunate poor sheltering in fetid slums, tens of thousands of tiny hopsack dwellings that spread across the city like patchwork blotches. In the same city Mukesh Ambani, a petrochemical multi-billionaire, is building the world's most opulent home, a billion-dollar two-hundred-yard-high twenty-seven-floor skyscraper. The towering glass palace will have several floors of open hanging gardens and enough room to house the family's six hundred servants.

On the way from the airport, each time the taxi stops at a red light, lame, blind and crippled beggars, both children and adults, scurry from the shadows to beg a few rupees. And each time I turn away, pretending not to hear the urgent rapping on the window by some poor waif. The money they collect goes straight into the pockets of the cruel men who run them like stables of animals sent into the street each day to earn their evil owners a comfy living.

I once heard a Catholic priest in the Philippines, a good man who had devoted decades to helping the downtrodden in that tragic country, claim in a sermon that the poor prefer to remain that way, that they never desire to be rich because they are the blessed of God. Perhaps he was just trying to make his flock's intolerable poverty more acceptable in a place where there is almost no hope that their children and their children's children will pull themselves out of the slums.

There lies Bombay's seductive appeal to hundreds of millions of Indians, in towns, hamlets and villages across the vast country. The heroes of Bollywood, India's gigantic movie industry based in Bombay, effortlessly make the leap from village hut to Bombay mansion, and millions have been lured to abandon their rural homes to settle here. Most eke out a meager living, enchanted by the parade of gleaming cars along the ritzy waterfront and richly clad party-goers sashaying into the grand hotels, proof that others have made it, and so can they.

I am content to settle down in a sleeping bag on the bare ground in the Australian outback, curl up on the deck of a steamer cutting through the Amazon or sleep on rocky slopes high in the Himalayas, but I am always partial to a few nights of five-star comfort along the way. The taxi pulls up by the Oberoi, one of Bombay's grandest hotels, perched by the waterfront along a two-mile curve of ocean named Marine Drive. The apartments here, strung along the beach like mouldy pearls on a fraying necklace, can be more costly, rupees per square foot, than the plushest apartments in New York and Tokyo. It is the free market at work: mass-producing manufacturers selling their products to a billion potential customers are competitive bidders for the limited number of apartments in Bombay's best and most spectacular addresses. A tatty apartment here can cost several million dollars.

Bollywood directors sometimes use the Oberoi's lobby as a set, their cameras lingering on the acres of floor marble, crystal chandeliers, the glint of gold-plated columns, the wide sweeping staircases and the hotel staff in their starched uniforms fussing around important guests. As I head for the reception desk, plump Indian businessmen sweep by, their triple chins and paunches wobbling in a deliberate display of self-importance, many accompanied by their tall, willowy "daughters" or "secretaries."

Ahead of me at the desk, a middle-aged Indian from Allahabad, his fingers and wrists gleaming with gold rings and bracelets, books a room for himself and the girl who stands silently at his shoulder, a few inches taller than he. She is in her early twenties, and is clad in an emerald silk sari with gold edging that falls seductively about her curves. A necklace of rubies and diamonds rests imperiously on her long, slender, amber neck, and her round liquid black eyes throb with sexual ripeness. The rich indeed live lives different from you and me, I think, as he leads her to the elevator and whatever *Kama Sutra* delights lie within their room—though his paunch must get in the way of the more interesting positions.

"How much is a standard room?" I ask the desk clerk. He looks like an upmarket butler, clad in a tasteful dark suit, regimental striped tie, and a cold-eyed look of smug superiority that seems to say that he could be playing polo or cricket at the club right now, but that he is checking in guests because it is a fun way to spend the afternoon.

He peers down his nose at me, as if I have come to repair the toilets. Perhaps it is the comfy way I dress on long flights, baggy tracksuit pants, a polo shirt, and road-stained Nikes. "Two hundred and fifty dollars a night," he sneers. The same amount would keep a poor family with a modest roof over their heads and in food for a year in Bombay.

"That's about ten thousand rupees. But you just gave a room to the Indian gentleman in front of me for five thousand rupees."

His eyes roll in derision. "It's Indian government regulations that room rates for foreigners are quoted in American dollars."

"But American dollars or rupees, it should work out the same, not double. From the look of the gentleman with his . . . daughter . . . he could buy and sell my country a couple of times over. Giving him a cheaper rate because of his race is discrimination."

"It's not. It's regulations. Now, do you want the room or not?" he snorts. "We're almost booked out and if you don't want it, someone else will."

Bombay is a hard place to find hotel accommodation, and because of the demand even three star hotels with spotted linen and paper-thin walls can charge $150 a day. So, I hand over my passport and credit card. His eyes brighten. "You're Australian," he smiles. "I'll upgrade you to a superior room. You beat our cricket team last time we went to Australia, but we enjoy thrashing your team when they come here to play."

Cricket is a revered and widely practiced secular religion in India. Australia might be the current world champions at this ancient English game, carried to the subcontinent by British colonialists, but Indians from the prime minister to rickshaw pullers are always eager to debate the merits of both teams. Steve Waugh, the former Australian captain, is a fading hero at home, but in India he is a demigod.

☠

My room is on the sixteenth floor. In the corridor, a pair of khaki-uniformed paramilitary police, cradling submachine guns, hunch on stools pushed against the wall. "*Namaste*," I say, "hello" in Hindi, as I edge past their gun barrels. They remain silent, glaring at me as if I could be a terrorist about to bomb the hotel.

I settle in the room, a plush package of large soft bed, fluffy pillows, ornate desk, giant TV, marble bath and ocean view. I unpack my bag, stow my cameras and money in the safe-deposit box, enjoy a hot bath and change clothes. It is a routine I have performed hundreds of times across the globe. Always intrigued by any mystery, I go outside to find out why Tweedledum and Tweedledee are holding fort in the corridor.

"Are you guarding someone?" I ask with a smile, following an inflexible rule when in the vicinity of men with guns. Always smile, stay calm, never get rude or angry.

No answer.

"Is there someone important in 1608?" My smile widens.

No answer.

"If you tell me, I promise not tell anyone else."

The door to 1608 opens revealing a handsome mustachioed Indian in his mid-thirties, barefoot and clad in a cream linen shirt and dark woolen trousers. His hair and eyes are raven black. "Why are you questioning my men?" he asks in that sing-song accent most Indians are saddled with when they speak English, transposing Hindi's dizzy run along a musical scale to the monotone language of their former rulers.

"I'm an Australian writer, interested to know if someone important is in the room."

"You're Australian?" he smiles. "Welcome to India, mate. We're going to beat you in the next cricket series, you know that, don't you? We'll knock over your best batsmen for a duck. Come inside and join me in a cup of tea." (A duck is when a cricket batsman gets out without scoring a run.)

He extends his hand. "I'm Ujjwal Nikam, India's Elliot Ness. I've prosecuted six hundred murder trials all over the state, and four hundred and seventeen of the accused got life sentences. My role is to battle terrorists and the underworld."

Phew, that is one of the most amazing introductions I have heard in three decades on the road. How can I impress him? I grip Ujjwal's hand. "Paul Raffaele, adventurer. I've lived with Stone Age people and pygmies, and am on my way to Benares to seek out sadhus who eat human flesh."

"Ah, the Aghoris. Yes, you'll find them in Benares, but be careful. They live by the cremation ghats and commune with demons and witches." His eyes twinkle as he takes from the cupboard an electric kettle and fills it with water. As it boils with a catarrhic rumble he tells me that a detachment of paramilitary police guards him twenty-four hours a day. "I move through Maharashtra state bringing the baddies to justice. My police are always with me."

Ujjwal points to the files strewn across the desk. "Do you know about the twelfth March in 1993 when Muslim terrorists exploded ten large bombs across Bombay, destroying buildings including the stock exchange. They killed three hundred and seventeen innocent people, and injured seven hundred and thirteen."

"Yes, I read that it was the Muslims' revenge for the massacre of hundreds of their people in Bombay in riots after Hindus destroyed a mosque the year before."

"Correct. Mobs of Hindus rampaged through the streets picking out Muslims and burning them in their houses and shops. But you can't justify one evil by perpetrating another. We quickly captured the Muslim terrorists who carried out the bombings, and we're nearing the end of their trial. I'm the prosecutor."

Caught up in the affair is Sanjay Dutt, a deceptively sleepy-eyed hulking thug and Bollywood superstar who has appeared in more than one hundred Hindi movies, in most of them the star and in many as a gangster.

"Though his father was a movie star and then a Hindu member of parliament, his mother was a Muslim movie actress. Sanjay got into bad company, like many Bollywood stars, who believe their demigod status among our people places them above the law of the land. When he was shooting a movie in Dubai he unwisely had dinner with the bombing mastermind, a Muslim Mafia don from Bombay, Dawood Ibrahim, who'd fled there. The next year Dawood's gang members here in Bombay gave Sanjay an AK-56 assault rifle, claiming it was for his protection. It's illegal to own such a weapon in India, but the silly man accepted it and some ammunition." The weapon was from a cache smuggled into India to be used in the bombings.

(As I write this the cases against the alleged bombers and Sanjay have just been resolved. Ujjwal has added to his impressive hit list. Judge Pramod Dattaram Kode found 100 of the bombings' conspirators guilty and sentenced twelve to death. He gave Sanjay, now 48, six years of rigorous imprisonment. "Sir, I made a mistake fourteen years ago. Please grant me bail," Sanjay begged in court. "I'd like to appeal. I am not mentally prepared for jail as I thought you will grant me probation." Justice Kode replied with all the gravitas and bad scripting of a judge in a Bollywood epic: "Sanjay, I have just done my duty. Everyone makes mistakes, but the element of criminality in you is incurable.")

☠

Gangsters and Bollywood have a long association. Some of Bombay's most powerful gangland Godfathers have hired assassins to kill famous movie stars, producers and directors who refused to cut them in on profits from their epics. Perhaps the best known victim is movie director Rakesh Roshan, father of Hrithik, one of Bollywood's most popular young male stars. In 2000 a gunman ambushed him outside his office and shot him several times, but he survived. On an earlier visit to Bombay, when I asked Roshan to tell me why he was attacked, he answered with a shrug.

More troubling is the clear evidence from Ujjwal and others that the Godfathers were all Muslims. Although they are a minority in Bombay, about two million Muslims live here, mostly in enormous ghettos dotted around the city. Muslims and the majority Hindus rarely mix socially, and a murderous tension thickens the air in India's biggest city, always on the edge of a boilover. Fundamentalist Hindus are always looking for an excuse to massacre more Muslims.

The hatred and rivalry go back many centuries, and the resentment regularly bursts into mob violence. The Hindus have called India their land since long before 1,000 BC and have deep-rooted and enduring memories, never forgiving the long Muslim reign. Muslims, from central Asia, invaded India in the eleventh century, but reached their apex of power with the conquest of the subcontinent by the poet-warrior Mughals in the sixteenth century. The Mughals dotted India with their magnificent palaces and buildings, including the Taj Mahal and Delhi's Red Fort. The dynasty spawned a clutch of legendary emperors, among them Shah Jahan and Akbar the Great. The British forced the last Mughal emperor from the throne in 1857.

The Mughals generally were even-handed rulers, but their ostentatious palaces, and heavy taxes that fell largely on the Hindus to pay for the luxury of their courts, engendered a hatred that simmers to this day. It exploded in 1947, when the British, hurrying to quit the huge colony they had ruled in one form or another for three centuries, gave in to the Muslim leader Muhammad Ali Jinnah and agreed to split India in two with a new Muslim state in the northwest to be called Pakistan, "land of the pure." Outraged Hindu mobs attacked Muslims all over the country, and the Muslims fought back. Hundreds of thousands of children, women and men were massacred.

That hatred still throbs. In 1992 Hindu zealots destroyed a revered sixteenth-century mosque in the same state as Benares because an archaeological survey hinted that it had been built over the remnants of a Hindu temple. Muslims rose up in revenge, especially in Bombay, where street fighting and bombings killed over four hundred people. Hindus stormed into the Muslim ghettos raping, looting and slaughtering more than two thousand people. Later, Muslims, it is claimed, torched a train in Gujarat, a state north of Bombay, incinerating fifty eight Hindu pilgrims returning from the site of the destroyed mosque. Hindu street mobs retaliated by killing more than one thousand Muslims.

In Stone Age societies in New Guinea I had witnessed evidence of never-ending cycles of "payback," where a tribesman of one clan is killed, that clan avenges his death with the murder of a single enemy warrior, which prompts the need for the first clan to kill once more in revenge. In India this madness is replicated but on a horrific scale.

☠

The next day I leave for the holy city and the Ganges. At Bombay's glitzy domestic airport, Indians in designer clothing throng the waiting hall, many carrying laptops, a symbol of middle-class success here. Reality sets in at Benares's ramshackle airport. The taxis here are Hindustani Ambassadors—many brand new, but built from the blueprint of a 1950s British car, the Morris Oxford. The bumpy road to the city passes by fields where dhoti-clad men guide bullock plows. Swarms of people inhabit every town and village and huddle of huts on the way. It is mid-summer, 110 degrees Farenheit, and the taxi is not air-conditioned. Through the open window, the wind is as hot as the blast from a commercial furnace. It sucks the air from me, and leaves me a little dazed.

I first came to Benares as a youngster when it was uncrowded, the lack of bustle befitting a holy city. About one hundred thousand people lived here then, but the population has swelled to three million. Most live in the new city whose streets are thronged with people and cars.

The medieval city, nudging the Ganges at its holiest place, has hardly changed in three hundred years. Jarwaharlal Nehru, India's first prime minister, said that the Ganges, "is the river of India, beloved of her people, round which are entwined her memories, her hopes and fears, her songs of triumph, her victories and defeats. She has been the symbol of India's age-long culture and civilization, ever changing, ever-flowing, and yet the same Ganges."

Benares, the Hindu holy city, on the holy river the Ganges. Up to a million Hindu pilgrims come here each year to bathe in the holy waters. The river banks are crowded with Hindu temples.

The streets narrow close to the sacred river. Saffron-robed holy men stride along the crumbling footpaths, and barefoot barbers hunker by the roadside shaving devotees' heads, while holy cows loll in the shade of banyan and pipul trees, munching greens reverently placed before them by pilgrims. A massive bull squats in the middle of a street, and cars and motorbikes edge around his regal bulk. Hindus revere cows, and regard killing or even injuring one as a heinous crime.

There are three thousand Hindu shrines and temples in Benares, and at each corner, whitewashed shrines garlanded with flowers honor one or more of the teeming pantheon. A million pilgrims flock to Benares each year and all make for the holiest of holies, a three-mile stretch of the Ganges River. Evening approaches as I arrive, and bare-chested priests in white sarongs perform at water's edge the night prayer ritual or *aarti*. They ring bells, beat drums, play flutes and chant ancient Sanskrit hymns as they offer the sacred river fresh flowers, incense and pots of milk.

Even an hour after the Sun set below the rim of the Ganges, the nighttime heat is still enough to send me weak-kneed along the street overlooking the river. At a bookstore there, the city's historian emeritus, Rana P.B. Singh of Banaras Hindu University, tells me that Hindu priests have been performing the *aarti* at this very spot for thousands of years. "To Hindus Benares is the center of the Earth, and its original name was Kashi, City of Cosmic Light," he says. "When the Ganges fell from the heavens to Earth at the Creation, Lord Shiva lay down his matted hair here to tame the river's awesome power, and the city has been His ever since."

To place the city's antiquity in context, he shows me a book by the Reverend M. A. Sherring, a nineteenth-century British missionary and historian who lived here.

> Twenty five centuries ago, at the least, it was famous. When Babylon was struggling with Ninevah for supremacy, when Tyre was planting her colonies, when Athens was growing in strength, before Rome had become known, or Greece had contended with Persia, or Cyrus had added lustre to the Persian monarchy, or Nebuchadnezzar has captured Jerusalem, and the inhabitants of Judea had been carried into captivity, she (Benares) had already risen to greatness, if not to glory. While many cities and nations have fallen into decay and perished, her sun has never gone down: on the contrary, for long ages past it has shone with almost meridian splendour.

Benares's immense spiritual charisma even attracted the Buddha as a young man almost 2,600 years ago when he was still a Hindu. At Sarnath, ten miles from the ghats, the wide steep sets of steps at the edge of the old city that reach down like fingers to touch the Ganges, he gave his first ever sermon, the "Turning of the Wheel," and founded a religion that would sweep across the world.

I explain to Singh, the historian, a kindly looking man in his late sixties, that I have come to the city seeking holy cannibals. He tells me that there are cannibal sadhus at the cremation ghats just a mile or so down the river. And, he admits that for more than two decades he was an Aghor devotee. "When I was a boy I got very sick and my father prayed to an Aghor saint seeking a cure. He promised that if I were made well then he would raise me as an Aghori. When I was in my twenties, I turned away from it. I just couldn't accept some of the nightmarish rituals that their sadhus perform. I now don't even believe in a God."

I return to my hotel, and the manager's face clouds when I tell him I am going down to the cremation ghats at midnight to try to find an Aghor sadhu. "Be careful, because the Aghoris will use the ghosts and demons that haunt the cremation ghats late at night to take control of your mind," he warns. "You'll later wake up down there not knowing what they've made you do, maybe even eating human flesh like them. Take a holy picture in your pocket for protection, and a bottle of whiskey. They like to drink a lot of alcohol; they see it as a sign of their saintliness."

CHAPTER 8

Clutching a six-dollar bottle of premium Indian whiskey I head for the cremation ghats in a *tuk-tuk*, a three-wheeler taxi whose canopy of rusting tin and rotting canvas rests uneasily upon a narrow, rickety chassis. The vehicle looks as if it would collapse if I gave it a stern kick. By day and well into the night Benares's streets are jammed with these ugly and not very efficient conveyances, belching foul smoke as their drivers tear along the crowded potholed roads, jousting for position like Formula 1 stars. But nearing midnight the streets are empty, and the *tuk-tuk* makes its solitary progress through a narrow maze of ancient passageways. Even the sacred cows are slumped asleep on the pavements, their massive heads resting against the crumbling concrete.

To reach the cremation ghats, the *tuk-tuk* driver heads deep into the old city and then sweeps back toward the Ganges. Thirty minutes later he drops me by the entrance to a darkened lane. "Follow it, and you'll reach the steps to the cremation ghat," he says.

The alleyway weaves through an ancient maze of houses. It is narrow enough that I can sometimes touch the stone walls of the towering residences on both sides simultaneously. They stink of dirt, grime and curry fumes accumulated layer by layer over several centuries. It's a musty acrid odor that I will come to know well during my two weeks in Benares. The only sound comes from my footsteps. But I am not alone. Ahead, silhouettes of men silently flit in and out of the shadows, appearing and disappearing around the many twists and turns.

Benares's swarms of tourists and pilgrims are always at risk of falling prey to the city's thousands of robbers and petty thieves, known in India as *goondas* and thugs. Almost alone in the alleyways I am an easy target, though I am deliberately carrying just enough money to get me back to the hotel by *tuk-tuk*. Sometimes this is wise and sometimes it is stupid because there are places where robbers will paint your head with your own blood for leaving your money back in the hotel safe.

Suddenly, I hear the thump of feet and for a moment my heart seizes in fear. *Goondas*! But then comes the tinkle of hand bells, and I relax—what thief announces his presence in such a way? Four men approach carrying what looks like a body tied onto a bamboo stretcher, and I press against a crumbling wall to let them pass.

The body's feet stick out at one end while at the other the head is covered with gold cloth. Garlands of marigolds are scattered over the body, which is wrapped in a red sari. Like me, the pallbearers are making for the oldest cremation ghat in Benares, Harishchandra, named after a mythical king, a forefather of one of the Hindu trinity, Rama. This king so honored honesty that he gave up his crown, his family and his freedom to repay a debt, and came to live at this very cremation ghat and toil alongside the untouchable caste that burn the bodies and dispose of the ashes.

The pallbearers do not seem to see me as they pass by, chanting *Ram Nam Satya Hai*, or God of Rama, the Supreme, imploring for the deceased his mercy. I follow them to the top of a set of steps that are almost hidden by the gloom. Clouds swathe the moon, and even when my eyes adjust it is still hard to see anything. But the eerie crimson glow of a body burning below at the ghat guides me toward the place where I hope to find a cannibal holy man.

With sweat streaming down my face, I edge down forty steep steps to reach the ghat and the Ganges, dark and threatening in the night. Looming over the river are the gloomy silhouettes of ancient Hindu temples. On a platform overlooking the burning body two men sit by an open hearth. Three well-fed dogs sprawl by the embers from a couple of glowing logs. A triton, a three-pronged spear brandished by Indian holy men, stands upright in the fire's ash. A bare-chested middle-aged man with neatly cropped hair sits cross-legged by the fire, and is clad in a black cotton sarong tied at the waist. He is chanting a Hindu hymn, his deep voice floating out over the holy river. A red prayer mark adorns his forehead. By him sits a short

One of the cremation ghats at Benares on the Ganges river where Hindus are cremated after death. One body is still burning on a pyre while fresh wood is stacked for the day's cremations.

wiry man with a rodent's sharp calculating features. He has on a dhoti, a length of white cotton wrapped around the waist and tied between the legs. He challenges me. "I'm Apu Chowdhury. Who are you?"

I explain that I have come to the cremation ghat to seek an Aghor holy man. Apu points to his companion. "He's Anil Ram Baba, the holiest Aghor saint at the ghats." I had expected to meet someone with dreadlocked hair piled high atop a face striped with sacred colors; eyes that burn with incandescent fervor; strings of sacred beads hung about the neck; a long beard that nestles in the crotch and a gaunt body that is naked or partly covered by a saffron robe. But the Aghor sadhu is clearly well fed, and his eyes hum with intelligence rather than blaze with mindless devotion. He looks as if he might be a banker come to seek absolution of his sins in the Ganges, and has stripped down for a midnight plunge into the holy river.

Apu seems to play the role of Sancho Panza to the Aghori's Don Quixote. He spies the bottle of whiskey, takes it from me and nudges the saint. Anil Ram Baba's eyes gleam when he sees the bottle. He reaches down to a bundle by his side to uncover the top half of a human skull wrapped in

Anil Ram Baba, the Aghor holy man at Benares on the Ganges. He lives at a cremation ghat, but took the author onto the Ganges in a row boat (see the river bank in the background) because he didn't want anyone listening to the conversation about Aghor chants. Here, he reads a Hindu holy book.

cloth. "This skull and the triton are my only possessions," Baba says as he fills the skull to the brim with whiskey.

"The skull is called a *kapalik,* and Baba drinks from it as part of his Aghor rituals," Apu explains.

"Drinking a full bottle of whiskey is part of an Aghor saint's rituals?'

Apu nods. "The Aghoris believe in practicing the most base acts possible as a sign of their holiness, and drinking whiskey is nothing compared to some of their other acts. They do what most other people could never dare do."

"Like eating human flesh?"

Apu gives me a knowing look. "Baba should tell you himself."

This hard-living saint throws the liquor down his throat in one toss, and fills the half-skull once more. He drinks it just as quickly, and turns to face me. "The logs came from the pyre of a body that's just been cremated down there," he tells me. "Apu got them for me."

He presses his thumb into the warm ashes and marks my forehead in the same way as a Catholic priest places ashes on believers' foreheads during the Ash Wednesday ritual. The priest's refrain runs through my mind. "From ashes to ashes, dust to dust."

"Put out your tongue," Baba commands. He takes more ashes on his thumb and goes to place them on my tongue, but I close my mouth and turn aside. I have no intention of putting a dead person's ashes on my tongue, even if it offends him. I do not know whether such an act would condemn me as a cannibal of sorts, and do not intend to put it to the test. Baba smiles, but says nothing. He must be long reconciled to the distaste others feel for the Aghoris' bizarre rituals.

By now the pallbearers have placed their corpse by the water, next to a bed of logs about waist-high. A man in his thirties, head shaven and clad in a white toga revealing a bare shoulder and arms, comes down the steps gripping wisps of burning straw, a child by his side. Neither had been with the pallbearers "It's the sacred flame," Apu whispers. "The dead person is probably the man's wife, the child's mother," he murmurs. "See, the body is wrapped in a red sari. If it were a man, the body would be wrapped in a white sheet."

As the husband passes his face is taut with grief, but the little boy, about three years old, is sleepy-eyed and seems bewildered to be out so late. The husband passes the flame to one of the pallbearers and circles the body clockwise several times, each time pausing by the head to pour water into the uncovered mouth, rigid in death. "It's a last blessing with Ganges' holy water," Baba tells me.

No women are present. Baba says they are banned from cremation ceremonies: he tells me their hearts are too weak and they cry too much. "It's bad luck for anyone to cry while the body is being cremated."

"Why?"

"I don't know. It's our tradition. The ban on women also comes from the time when we Hindus practiced *sati*, when a woman would join her dead husband on the pyre and be burned with him. Sometimes the family forced her onto the pyre, but usually she went willingly. That gave her great blessings when she was reborn. When the British banned *sati* about two hundred years ago, families discouraged their women from attending the cremation of their husbands in case they threw themselves onto the fire and got the family in trouble with the colonial government."

Baba believes that was a change for the worse. *Sati*, practiced for millennia, "was the supreme demonstration of a woman's love for her husband," and the shrines throughout the city to women who immolated themselves because of love or a reverence for religious tradition or under compulsion are still venerated. Men never practiced *sati*. Baba says, "Women are replaceable, men are not. A man has to earn money for the family to live, but a woman stays at home. A man can marry several wives, but a woman must dedicate herself to just one man."

Apu interrupts, touching my shoulder and pointing to the pyre. The husband has retrieved the flame and sets the pyre alight. "It takes about eight hundred pounds of wood to form the bed for the cremation," he tells me. "The cost for ordinary wood is about 5,000 rupees," about $125, "but the rich pay about one million rupees to buy enough sandalwood for a cremation."

The wood has been slathered with ghee, clarified butter, and swiftly catches alight, casting eerie flickering shadows across the mourners. After circling the fire a few times the husband stands to the side gripping the child's hand, his face racked with sorrow. About thirty minutes into the cremation, with the fire crackling at full blaze, the skull has become visible at the top of the pyre, with most of the flesh burnt away, causing it to grin gruesomely. At the bottom the body's charred legs, now stripped of most of their flesh, jut out from the pyre.

The husband takes a bamboo pole and hits the skull, breaking it open. "That releases the soul from the body so that it can journey into the sky and get ready for rebirth," Baba says.

I wince, glad that I am not a Hindu. I doubt I would have had the courage to do the same when my father died some years ago. Perhaps, long ago, Hindu high priests devised this grisly death ceremony to hammer home the core belief that the soul is immortal, but the body bearing it is just a lump of meat and bones. Once it is separated from the soul, it is no different from the carcass of any animal you might find in the marketplace or slumped dead in the forest.

The sacred flame is kept by the Doms, a hereditary caste charged with tending the cremation ghats. Doms like Apu believe that thousands of years ago Shiva gave them a sacred flame, which must be used to ignite all Hindus pyres, or else they cannot be reborn.

I tell him it all sounds self-serving. Apu chuckles, more to himself than at me. "Our lineage goes back, generation by generation, four thousand years. How far can you trace yours back?"

"About one hundred and fifty years."

"That's like a spit against the ocean." He points up the steps. "We keep the sacred flame in a shrine at the top. Here, at Harishchandra, it has burned for many hundreds of years without ever going out."

"And before that?"

"We Doms are sure it has been here for at least four thousand years."

"The same flame?"

"Yes, the same flame."

"What happens if it did go out; say, if a Muslim came by and threw water on it."

Apu's eyes turn hard. "That would be a declaration of war. Hindus all over the country would rise up and teach the Muslims a lesson they'd never forget."

Young boys of the Dom caste, who tend the cremation pyres, search in the water after the remnants of the cremation have been shoveled into the Ganges. They are looking for coins, jewelry and any other valuables.

"But you'd still have no sacred flame."

"All over India at the cremation ghats we Doms keep the sacred flame which originally came from Benares. So, we'd just go to the cremation ground nearest Benares and bring the flame back here." In any case, the Doms' vigilance would always prevent the flame being extinguished.

He promises to introduce me to his cousin who lives not far away downriver, Jagneesh, "King" of the Doms. Apu believes Jagneesh's command of the sacred flame at all of India's cremation ghats makes him the most powerful person in the country.

Apu is a Dom, but not an Aghori. He met Baba because the sadhu lives by the cremation ghat where Apu toils during the day. As we talk, Baba grips the triton and murmurs prayers. When he finishes, the holy man explains that his chants chase demons from the ghat. They come at midnight, and he says that if he doesn't ward them off with the chants then they could harm me.

I mention my hotel manager's warning about ghosts, prompting Baba to peer at me with considerable sympathy, as if I were a simpleton. "In Hinduism, we don't believe in ghosts. Once a person dies his soul goes immediately to the place where it waits until it finds out what it will be in the next life. So, how can there be ghosts here?"

One of the fattened dogs lifts its head, growls, leaps to its feet and races into the darkness, followed by the other two. From the shadows another gang of dogs has come to challenge, and for a few minutes the dogs hurtle up and down the steps, growling, snarling, barking and baring teeth. But the battle depends more on bluff than bite. Having seen off the challenge without a single injury, the dogs return to the hearth and slump with gleaming eyes by the fire. "They eat well here," Baba say, "so the other dogs are always trying to dislodge them."

The dogs' staple food are corpses. "A body burns for about three hours," Apu says, but sometimes the wood used is not enough and some charred flesh remains. "With women, it's usually the buttocks, with men the flesh on the chest. Doms push all the remains into the shallows, ashes, wood, bones and whatever flesh is left. The dogs snap up the flesh, and grow fat on it."

Baba says that the families of the cremated are not alarmed by this gruesome end to their loved ones. He tells me that to devotees, "Once the soul has gone from the body, it's just a lump of flesh; and if it can feed another creature, then the family believes it is a blessing on them."

"I'm told you sometimes eat leftover flesh from the pyre as part of your Aghor ritual."

Baba moves his head from side to side, a gesture that almost anywhere else in the world means no, but in India is a sign of agreement. "I've spoken enough and must go back to my prayers. Come back in a couple of days, at midnight, and I'll tell you more."

"Should I bring another bottle of whiskey?"

Baba's eyes shine. "As you wish," he murmurs.

CHAPTER 9

I wake at 4:30 the same morning. Breakfast is a bowl of curds scooped up with unleavened bread, and my companion is a chapter from a book, *Banaras: City of Light*, written by Diana L. Eck, a Harvard professor specializing in India. She wrote:

> If we could imagine the silent Acropolis and the Agora of Athens still alive with the intellectual, cultural, and ritual traditions of classical Greece, we might glimpse the tenacity of the life of Kashi (Benares). Today, Peking, Athens, and Jerusalem are moved by a very different ethos from that which moved them in ancient times, but Kashi is not.

Just after five, in a ritual that could have occurred with scant difference two thousand years earlier, I join hundreds of pilgrims murmuring salutations to the god Rama and dodging the holy cows as they swarm down a cobble-stoned path toward the Ganges, about one hundred paces from my hotel. Sunrise is at least an hour away, but from below the horizon the sun's unseen rays have tinged the clouds above a faint shade of pink.

The women pilgrims are clad in saris that cover most of their bodies but leave their stomachs and arms bare, the men in white dhotis, and many of the children in their Sunday best, pants, coats and dresses, and polished shoes. We pass through a gauntlet of sadhus who look just as I had expected. They are bone thin, dread-locked, wild-eyed, almost naked, gaunt, bearded, and painted with sacred colors and paste to signify the fire blazing from the god Shiva's third eye. Many grip tritons, like holy warriors. Indian has millions of sadhus, ascetics who renounce all worldly desires and live off pilgrims' charity. Some press chillums, small pipes, to their noises, drawing in gusts of *ganja*, or cannabis, a drug considered a gift from the divine, bestowing on the holy men mystic communion with the Hindu gods.

Men, women and children bend to have the sadhus affix red prayer marks on their foreheads, and then drop a rupee or two in the holy men's bowls. Barbers squat by the path using cut-throat razors to shave bald the heads of male devotees, a ritual cleansing before they plunge into the holy Ganges.

At the river, boatmen beseech the pilgrims for custom. I am with Dharmendra Tiwari, an experienced guide, and he hires a rowing boat for five dollars an hour. Tiwari is a stern-faced man with a moustache and topknot tied with saffron twine. He sees me looking at it. "We Hindus believe that energy flows up through the body to the head where it can escape, and so we tie the hair at the back to keep the energy safe inside us," he says.

The sky stretches across the heavens like a bolt of raw pearly silk. The air has lost its sting overnight and is refreshingly cool as we push off and head along the banks of the three-mile stretch of the Ganges that nudges the old city of Benares.

"Benares suffered several major destructions by the Muslims, from the twelfth to the seventeenth centuries, and the invaders sometimes reduced the city to rubble, but our Hindu ancestors always rebuilt the city," Tiwari tells me as the boatman bends his back to the task, moving us swiftly down the river with the help of a sturdy current. "Once the British took power, in the nineteenth century, Benares became the second home for rich aristocratic families from all over India. The Rajahs and Maharajahs came from Rajasthan, Mysore, South India, Gujarat, and many more regions. They built those palaces you now see, making Benares India's first cosmopolitan city. The rulers of the many princely states constantly fought each other, but here along the Ganges they often had palaces next door to each other and would meet and parlay before heading home and resuming war."

Little has changed since Mark Twain took a boat along the riverfront in 1896. In his book, *Following the Equator*, published the following year, he wrote: "Its tall bluffs are solidly caked from water to summit along a stretch of three miles, with a splendid jumble of massive and picturesque masonry, a bewildering and beautiful confusion of stone platforms, temples, majestic palaces softening away into the distance. And there is movement, motion, human life everywhere, and brilliantly costumed—streaming in rainbows up and down the lofty stairways, and massed in metaphorical flower-gardens on the miles of great crammed platforms at the river's edge."

The ghats are a microcosm of India. Bengali traders, Rajasthani camel herders, Gujarati railway workers, Bombay movie stars, Tamil fisherfolk, and people belonging to hundreds more tribes, castes and sub-castes flock here—more than 60,000 a day—for a splash of salvation. They flaunt their region's ethnic dress, the many colored turbans, saris and tunics on the way to the ghats, but along the riverfront, the pilgrims cast off most of their clothes and join their palms reverently as they slip into the Ganges to dip their heads repeatedly beneath the murky water to wash away their sins. It sometimes lands them with diseases such as cholera, hepatitis and typhoid. "They come from India's far corners for the ritual bath in the Ganges," says Tiwari, "and don't worry about the dirty water."

On the bluff, in a medieval tableau, soar crumbling temples and palaces pushed against each other, like an encyclopedia of the country's architecture. India at a glance. Most are deserted, over-run by aggressive bands of monkeys whose clans have lived here for centuries. Bells, drums, flutes and conch-shells echo without pause along the river because this is the kingdom of the *pandas*, the Brahmin priesthood. Wearing white dhotis and smeared with sandalwood paste on the forehead, hundreds of priests clamor around the pilgrims, offering for a price to bless them with prayers to Shiva and others in the vast Hindu pantheon.

Sadhus plastered with sacred clay sit beneath enormous umbrellas on the many small circular platforms overlooking the river. Although dawn, the most sacred moment of the day, is yet to come, Hindus are devoutly washing and praying in the Ganges. Tiwari tells me that many will consult the sadhus on questions of religion or astrology, and seek their blessing, before heading for one of the hundreds of shrines dedicated to Shiva, to empty a copper jar filled with the holy river water on a representation of the god's *linga*, or sacred penis.

In each shrine an erect squat pole of darkened stone, Shiva's life-giving *linga*, rises from a sweep of flat stone. Its raised edges almost complete a circle and then dip into a giant thick-lipped V, the shape resembling the *yoni* or vagina of Shiva's wife Parvati. "Each day devotees pour milk and

Ganges water on Shiva's *linga* in the many temples here, and then coat it with sweet-smelling sandalwood paste," Tiwari says. "There are more than three thousand *lingas* of Shiva in Benares." This does not surprise me in the land of the *Kama Sutra*.

Shiva, the God of Destruction, is one of the Hindu trinity, along with Brahma the Creator and Vishnu the Preserver. Each is responsible for one of three states that make up what Singh, the historian I had met the day before, called the Hindu "frame of the cosmic reality." The three are evolution, existence and involution, and they occur cyclically and infinitely. "Shiva, being the last to complete the cycle from which a new cycle begins is known as Mahadeva, the Supreme Divinity, and Shiva's *linga* represents the unity of these three states of the cosmos."

I cannot think of any other religion where the penis of one of the most powerful deities is not only explicitly portrayed and glorified and pampered in the temples, shrines and holy books, but plays such a powerful role in the smooth running of our world. Shiva is not only the patron god of Benares but also the most popular of the three supreme deities here. As a Benares city guidebook explains, "Shiva is not perfect: he is easy to anger, easier to please, prone to impulsive mistakes. It is because of this that he is loved so deeply by his devotees. And to seek his blessings you have come to his special city. Here, Shiva is everywhere."

According to Tiwari, Benares has many names, and one is 'the home of Shiva.' According to Hindu tradition, Brahma weighed Kashi—the concentration of cosmic light—against the heavens. Kashi sank, and the heavens, the home of the gods, rose. Vishnu positioned his legs on either side of the city, forming the Varana and Asi rivers, which bind the old city north and south and protect Benares from evil. Shiva came here to live, ruling the cremation ghats.

But even if he is a favorite god, Shiva is not alone, Tiwari says—Hindus have thirty million gods and goddesses to venerate. "Thirty million? It's hard to believe you have even thirty thousand gods and goddesses. That's just over eight for each day of the year."

"Would you like me to name them all?" Tiwari counters. "Bhimeshwar . . . Pavarti . . . Ganesh . . . Krishna . . . Kali . . . Saraswati . . ."

"Enough," I plead. I make a quick calculation in my notebook. "If it takes a second for each name, we'll be here for 347 days, or almost a year, and that's without going to sleep."

Tiwari gives me a smug look. "You Christians have only got three gods, and the Muslims only have one. I don't know how you cope." Not wishing to kick off a religious war between us, I let his comment drift into the wind.

This teeming divinity does provide for a god for every cause. Geoffrey Moorhouse, in *Calcutta: The City Revealed*, wrote that Hinduism

> enshrines a bewildering pantheon of figures who together are venerated for every conceivable reflex and incident in the human condition and philosophy. There are gods as jolly-looking as Ganesh, sitting comfortably with his elephant's head, who is invoked by writers to bring them success. There are goddesses as elegant as Sarasvati, riding upon her gorgeous peacock, patron of music and inventor of Sanskrit. And there are scores of godlings with more unfortunate connotations like Manasa, who is worshipped in Bengal, as an antidote to snake bites, and Sitala, who is particularly idolized by people along the Hooghly (a river that passes through Calcutta) during outbreaks of smallpox.

A splash brings me back to the Ganges. Tiwari has leaned over the side of the boat to cup water in his hand and now sprinkles it over his head and drinks a few drops. The water of the Ganges, the ultimate receptacle for the city's hundreds of thousands of toilets, is among the world's most polluted, and I admire Tiwari's bravery.

A level of 550 fecal coliform bacteria per hundred milliliters is believed to be safe for humans bathing in India's rivers, using the standard set by the World Health Organization. At Benares the level can be up to three thousand times higher, according to Veer Bhadra Mishra, a retired water-source professor and the revered hereditary *mohan*, the high priest of the temple of the monkey god Hanuman which overlooks the Ganges. "The government pumps the city's raw sewerage straight into the river, millions of litres each day, and allows corpses to float into it," Mishra would later tell me. "I belong to the Ganges, I talk to her all the time, she is my mother, I tell her all my happiness and problems. I want to touch her, to submerge my body in her every morning when I wake. My ancestors drank from the Ganges, and she provides 40 percent of fresh water for India's people. But the Ganges is so polluted here that I no longer drink from her."

Hindu artists paint the Ganges as a beautiful goddess who sits cross-

The pilgrims bathe in the filthy waters, which have up to three thousand times the accepted WHO limit for fecal matter in the water. The latrines of Benares are emptied into the river.

legged in perfect repose on the back of a peaceful crocodile and grips a lotus flower. To cope with the pollution one needs to fix that image firmly in the mind when coming anywhere near these malodorous waters.

As our boat passes Harishchandra ghat, the sacred flames are consuming two bodies on beds of burning wood. One body is still hidden, but at the other flames lick at a charred head, legs and an arm raised high, as if giving the world a farewell salute. A skinny-legged Dom in a white dhoti pushes the remains of another cremation into the Ganges while a water buffalo looks on with bovine indifference as two boys stride into the shallows and begin sloshing the murky water in wickerwork baskets, as if panning for gold. "They're Doms, untouchables," Tiwari says. "They're going through the ashes seeking any jewelry or coins left on the body."

A pair of dogs spy something in the leftovers and race down the steps to the water's edge. One snaps up a tidbit about the size of a melon, and the other grabs the other end. Snarling and growling they pull at it until one wins the tug of war and scurries away in triumph. The losing dog howls in anguish.

"What were they fighting over?" I ask Tiwari.

"You wouldn't want to know. As a Westerner I don't think you'd understand."

"Let me try. Was it some flesh from the body?"

Tiwari nods, closes his eyes and begins chanting a prayer.

I look around for Baba, but he had told me he prays until 4 am by the cremation ghat and then sleeps until the afternoon. A concrete hut sits by the cremation grounds. "A very holy Aghor saint has lived there for years," Tiwari tells me when he had finished the prayer. "He comes out at night to pray to the demons, and even eats human flesh. Don't ever come here in the dark, because he can be dangerous. As you've asked to meet an Aghor sadhu, I'm taking you to another cremation ghat where a holy man lives who is not as powerful as the sadhu over there and is less likely to harm you."

I see no reason to tell him about my meeting with Anil Ram Baba just a few hours earlier. Tiwari points downriver to a mansion high on the bluff and guarded at each end by a life-size statue of a tiger. "That's the home of the Dom king," he says. "He's one of the most powerful men in India."

A mile further down the Ganges we alight at Manikarnika ghat. It is the biggest cremation ground; up to two hundred bodies are consumed here by the sacred flame each day. At this early hour, corpse-laden biers are lined up along the steps leading to the pyres. Dead men are swaddled from head to toe in white cloth, women in red silk. Doms, wiry men with heavy moustaches and white dhotis tied between the legs, swarm about the ghat selling logs for the pyres, building the wood into what they call beds for the corpses, and bearing the sacred flame down to the river on wisps of straw or twigs. Smoke from the pyres fogs the steps that lead up from the river.

Singh had told me that up to thirty-eight thousand bodies are cremated in Benares each year, their owners seeking *moksha,* or enlightenment. "Many come to spend their last days or even years on Earth to ensure they're cremated in Benares when they die."

Huge piles of logs are stacked on the steps by the water, ready for more pyres, while barges piled high with logs are tied to the riverbank. It looks as if it is going to be a busy day for the Doms. Just a few yards downriver from the pyres dozens of pilgrims—children, women and men—have entered the water and are dipping their heads beneath its surface, which has been stained the color of gun metal by the ashes of the cremated faithful.

Looming over the ghat are a handful of small but highly sacred temples, their narrow spires painted a color that seems in one kind of light to be pastel pink, in another to be a soft shade of brown. One temple honors Vishnu, and the god's footprints where Hindus claim he once strode this very spot have been encased in a slab of marble. Another temple holds a revered *linga* of Shiva, and from its inner sanctum comes the clang of cymbals and chant of priests from the Brahmin caste, the supreme Hindu caste. "I'm a Brahmin," Tiwari reveals, the highest caste, "and it is our sacred duty to protect these shrines from the Muslims."

Close by, on a platform overlooking the Ganges, a bare-chested long-haired sadhu clad in a saffron sarong and a profusion of neck beads sits in the lotus position with crossed legs. His hands are joined, and he seems to be praying. "That's Shankar Giri, an Aghor sadhu who lives on the platform," Tiwari says. "Let's go and talk with him."

☠

The American academic Ron Barrett had told me from Stanford that Hindus are under the command of a strict set of practices that set the limits of ritual purity and pollution, performing daily the former and banning the latter. That is why caste Hindus avoid contact with untouchables because they are considered impure. "Aghor takes you beyond the restrictions of these practices of ritual pollution and purity so that its practitioners don't have any aversions toward each other any more. It shatters all social hierarchies, and India is built on religious and social separation by caste, so that all people are equal. It's a belief that liberates you. That's what attracts me to Aghor."

Living by a cremation ghat is ideal for the Aghor sadhus. It allows them access to human flesh when they desire, the chance to use mantras to battle the demons that swarm about the cremation places at night and the opportunity to smear themselves with ash from the cremation pyre. This is a graphic demonstration of their desire to reject any adherence to the mainstream Hindu concepts of pollution and purity.

Shankar's platform near Manikarnika ghat is eight steps high and about two yards square, and he offers me a greeting with joined hands as he

An Aghor sadhu, Shankar Giri, who lives near a cremation ghat at Benares on the Ganges river. He is here twenty-four hours a day and took the author for a dip in the Ganges.

murmurs, "Namaste." I return the traditional Hindu salutation and sit down beside him on the platform. I notice that one set of his beads consists of about fifty miniature human skulls carved from white stone and strung along a necklace.

I have brought along a bottle of whiskey. As Baba had done, Shankar removes the top half of a human skull, his *kapalik*, from a cloth bundle and pours the liquor into it, filling it to the brim. Before he drinks he withdraws what he says is a human finger bone from the same bundle, dips it into the whiskey and, as an offering, flicks the liquid onto a small holy picture in a frame. It shows a son of the gruesome Kali, standing on the body of a be- headed enemy and holding aloft the bloodstained severed head.

"Kali-Ma is my mother, and her son is my brother," Shankar tells me. He turns to see if the sunrise is near. Its powerful rays have begun streaming over the horizon, and he stands up. "We'll talk later because I have to say my *puja*, my prayers, as the sun rises. Join me."

It sounds more like a command than an invitation. I had sworn not to enter the putrid Ganges, fearful of picking up cholera or typhoid or any of the many other diseases present in the sickly water. But if I do not join the Aghor holy man, he might refuse to talk to me. I decide to just immerse myself at waist-level, hoping to come to no harm.

Shankar slips easily into the water waist-high just as the tip of the sun pokes over the horizon. He waits until its rays stream toward the riverbank ghats, and then holds aloft a handful of Ganges water and lets it fall over his head. "The sun's rays pass through the water and empower it," he tells me. "The water then cures any illnesses I might have." He ducks his head under the water, immersing himself fully in the river several times, his lips moving as he offers what he later tells me was silent prayer beseeching Shiva to bless all the people in the world.

The sadhu beckons me to enter the Ganges. I tread carefully down the steps, determined not to immerse my head in the slimy water. But I fail to see that the final step, underwater, is coated with moss. The moment my foot touches it, I slip and plunge head first into the Ganges, and then rise to waist-level, spitting out water that rushed into my mouth. Tiwari smiles, and I laugh at my comical misfortune, but the sadhu is too focused on his own prayers to see the slapstick humor in my clumsiness.

Back on the platform the sadhu and I let the sun dry our bodies as he takes my palm and peers at it. He lets out a soft "ahh." "You're a beloved of Shiva," he murmurs, more to himself than to me. "Shiva will strike down anyone who stands against you."

He shows my palm to Tivari. "Look, the marks of Lord Shiva. I've only ever seen one other palm like this, and I've looked at thousands of hands."

He points to three intersecting lines at the base of my palm which clearly resemble the prongs of a triton. He then follows one of the lines up my palm to the midway point where a clear X connects two parallel lines that cut across the palm forming a clear outline of a hand drum etched into the flesh.

Tiwari looks impressed. "The triton and the hand drum are the two symbols of Lord Shiva," he tells me. "Shiva has marked you as a favorite."

"Only if I believe in Hinduism," I counter.

"This should be enough reason for you to do so," he replies.

Shankar begins murmuring prayers as he forms shapes with his hands and fingers, two of them I recognize as a *linga* and *yoni*. "It's an Aghor ritual," Tiwari says. "Many of the shapes he's forming are secret, and have meaning only to other Aghor sadhus."

After about ten minutes the sadhu dips his thumb into a small container and marks my forehead with a red prayer mark. Then, he takes a string of red beads from his neck and places them around mine. "These were given to me by my master," he says. "I give them to you because you are a child of Lord Shiva." I am thankful he did not choose the miniature skulls, and decide to wear the beads for the remainder of the time I am in Benares.

As monkeys boldly scamper along the steps above us, we talk about how he became an Aghor holy man. "I'm thirty years old and have been an Aghor sadhu for seven years, but my home is sixty miles from here," he says. "When I was a novice my master summoned me to Benares during a time of flood, and on the riverbanks of the Ganges were many dead bodies. My master broke apart two skeletons and made a line of the bones along the ghat. He began to chant an Aghor mantra, and the bones jumped into the air and started fighting each other."

"Were you drinking whiskey at the time?"

Shankar ignores my joke, or perhaps he did not hear it. "Do you want me to stop the bones fighting?" my master asked. "I nodded because I was frightened as I didn't know the chants then to command the bones. My master chanted another mantra, and the bones fell back to the earth and were still. That's when I decided to follow my master and do whatever he told me. Now, I live on this platform day and night. Devotees bring me food, and sometimes alcohol. I watch the fires of the dead while chanting mantras, especially from midnight to four in the morning, when the sky here at the cremation ghat is filled with demons. I've collected more than one hundred human skulls because they give me special power, and keep them in a secret place."

"But why join the Aghoris?"

"Because they follow the left-hand path of Tantra, and I find that suits my nature. God is perfect, and so cannot create imperfection which is only an illusion of the human mind. To reach a holy state we Aghoris practice acts that most people mistakenly find repulsive, unholy, taboo."

"Is eating human flesh one of those acts?"

Shankar nods.

"Have you eaten human flesh?"

He nods again. "Just a few times, though when I am more practiced in chanting the proper mantras I expect to do it more often. Not long after I began my study and practice of the Aghor way, I obeyed my master's command to sit on corpses washed up on shore and to meditate. At first the smell was terrible, but soon enough the power of the mantra given to me by my master helped me not notice that. I also went to brothels and sat on the beds of prostitutes while they were with customers."

"They didn't mind you there?"

"They ignored me. I paid them from money given to me by devotees. I also sleep with prostitutes, for at that moment when we join, I am Shiva and she is his beloved Parvati. But I must always hold back my semen. About two years ago my master told me I was ready to eat human flesh, but I was still not strong enough in my mind to take flesh from a burned body at the ghat in front of other people. So, when I saw a corpse floating down the Ganges, I swam out to it. It must have come from way upriver because it was a whitish color. It was a woman, and so she might have been pregnant. Her body was too swollen for me to tell. Almost all Hindus are cremated when they die, but there are six kinds of people who are not cremated because they're considered already purified, and they're tied with a stone and dropped into the sacred river. They are children, lepers, sadhus, victims of cobra bites, those who've died from smallpox and pregnant women. I chanted a mantra my master taught me as I grabbed the body. The flesh was very soft and some of it came away in my hand. I kept chanting the mantra and ate it. Because of the mantra's power, the flesh tasted as if I were eating a mango, very delicious."

"But how does such an act make you more holy?"

"Because only those Aghoris who have practiced the mantras, and meditated, are strong enough in their minds to carry out such an act."

"That still doesn't explain your desire to become a cannibal."

"I'm not a cannibal. The person is already dead, and so the body is just a lump of flesh."

"You're still a cannibal because it's human flesh."

"You may think what you like," he replies. "Our holiest sadhu, Anil Ram Baba, lives at Harishchandra ghat. Ask him."

He turns away for a few moments and then swivels to face me, his eyes bulging with a manic stare. I stare back at him as if we are engaged in some kind of mental duel for a minute or so until he turns his fearsome glance onto the Ganges. "He was trying to dominate your mind," Tiwari says as I rise, judging this a good time to leave. "Then you would have been under his command. But as a child beloved by Lord Shiva, you were more powerful than him."

Those who know the Aghori accuse them of attempting to hypnotize people and have them do not very nice things. This is the moment when the Aghori Shankar Giri turned incandescent eyes on me in an attempt to overcome me mentally.

The author with
Shankar Giri.

Shankar reaches up to touch my arm. "Be careful of the many fakers along the river who claim to be Aghor sadhus, but who are imposters, here to make money from the pilgrims. One calls himself Black Bom Bom Baba. He sits beneath poles bearing human skulls. But at night he goes back to his wife and children, his pockets bulging with rupees from the pilgrims. True Aghor sadhus never marry. All the women of the world are our mothers and sisters."

The temperature has soared to about 110 degrees Farenheit, and after a curry lunch near my hotel I take a reviving siesta. Then I go to meet my new guru.

CHAPTER 10

The *tuk-tuk* stutters through the mid-afternoon traffic. It is gridlocked as usual with too many vehicles cramming roads built for a sprinkling of bullock carts and rickshaws. The coming of prosperity for a vast emerging middle class has spurred a lust for cars, but the government, its funds leached by nationwide corruption by officials at village, city, state and federal levels, puts little money into upgrading the road system from the narrow potholed macadam that makes up most major thoroughfares.

Indian society in many ways deliberately represses the individuality of its people, forcing them to conform with inherited caste and status. As if to assert the balance, drivers here enforce their presence with nonstop honking, demanding that a path be cleared for them through the jumble of vehicles even when it is obvious that the hold-up is ahead, out of sight. This makes any journey in a *tuk-tuk* during the busy hours a painful experience for the ears, and a supreme test of patience. Now and then I lose control and shout at the drivers of passing cars or motorbikes, but my voice is lost in the clamor of honking.

I have decided to take a course of four lessons in Tantra from a revered master, Anand Subrahmanyam Shastri. Happily, his house is in a quiet backstreet, surely an essential requirement for a philosopher who makes his living by thinking and teaching. He is a tall, spare bald man in his sixties, and is clad in a loose-fitting white sarong and white cotton shirt with no buttons. "Call me Swamiji," he says as I settle in a sofa in his study, which holds a small TV, a statue of the god Ganesh, a bookcase and a whiteboard.

"Tantra is a clear pathway to divine bliss, enlightenment," he tells me, "but it takes decades of practice to attain this. What I can do is show you the path, but it is up to you if you wish to follow it. There is another path, the Vedic, and this is followed by householders. The Tantric path is followed by what we call renunciators, with no wife, no children, no family, so that they are free of social obligations."

Swamiji was once a husband and father, but he renounced his family and forswore all desire. I am tempted to ask him why he needs the hundred dollar fee I am to pay him. Cynics might consider that could signal some remnant of desire polluting his purity of soul, but I don't ask. It would probably put my lessons in jeopardy, and he is the spiritual master of highly-placed disciples from all over India.

Swamiji goes to the whiteboard and draws a series of circles and squares that he says represent Earth-bound life and life eternal. "Our soul moves through many cycles of life, and what we are in the next life depends on how we acted in this life, until we reach *moksha* or enlightenment," he tells me. "Divine powers are controlled by special sound symbols. Mantra is the most powerful, it is naked energy, and Tantra is the method of using the mantra's power for fulfilment of ambition and enlightenment. Mantra, divine sound, is the nucleus of the universe."

He senses my disbelief. "Before creation, everything was still, there was no sun, moon, no life in space. Life comes from air, the essence of mantra, and was injected into space through air."

Swamiji's face turns grim. "I can kill by using a mantra, its sound energy causing the person to vomit blood and have a heart attack. I have never put it to the test, but many years ago I saw my master kill a much hated and feared criminal in this way just by using the same mantra. Blood poured from the criminal's mouth, he shook all over and then he died in front of my eyes. My master taught me that mantra, but I have never used it."

I thought a mantra was a kind of slogan, not a deadly weapon. Swamiji smiles at my disbelief. Taming a mantra's power is not easy, he says. "A master gives a mantra to his student who must chant it seven hundred thousand times over twenty-one days to achieve whatever he seeks. At the end of your lessons I'll give you a special mantra. I can see you are made for some great work, but I can't reveal it because it is up to you to find it."

These holy philosophers are well practiced at flattery, and so I perseverve with the lesson. A mantra can be short or long, depending on the purpose. He chants a drawn out syllable, AAAWWWWWMMMMM, the same sound as the Buddhist OM which he claims is the mantra that regulates the path of the planets in our universe. "There are no imperfections in the world," he says, echoing the claim of Shankar. "Imperfection is just a human illusion, something the Aghoris as Tantrics also believe."

But then he edges close to contradicting himself. "Male is virtuous and female is vice. If there is no vice in the world then there is no visibility in our world. Every entity on earth is a mixture of good and bad. If there is no darkness then there is no value in light. If there is no struggle between noble and evil forces then there is no life."

Over the next hour Swamiji leads me into a philosophical maze. He tells me that our world is controlled by five universal principles, headed by Shiva representing the masculine and Shakti the feminine. Knowledge, creation and the transcendental are the other categories in the divine bundle. Each of these is then subdivided into more numbered principles that divide as effortlessly and rampantly as amoeba in a puddle of water. Swamiji says this immense depth and width of knowledge is the Tantra way of reaching for the divine, but to me most of his explanations are hopelessly obscure, perhaps sometimes deliberately.

Swamiji returns to the circles and squares on the whiteboard. He calls them yantras, divinely inspired geometric diagrams that contain and intensify the essence and power of mantras. "A yantra's graphic nature allows a person to focus on the essential nature of the mantra," he says. He erases the board to draw another yantra, this one a precisely drawn jumble of triangles, squares, rectangles and circles. He then draws another, each subsequent one becoming more complicated and obtuse. He spends a few minutes drawing the most complicated, and beautiful, yantra I have seen. It is made up of several large triangles whose interlocking sides form forty-three much smaller triangles. Drawn around the triangles are two circles embroidered with lotus petals. Swamiji says they represent a powerful goddess and her courtiers.

Toward the end of the lesson, Swamiji reveals that for many years he was an Aghor devotee, and I tell him that I have come to Benares principally to meet Aghor sadhus and learn about their practices. "I was attracted as a young man because it seemed an exciting form of Hindu belief," he tells me. "The Aghoris revere five objects. Mantra, madya or wine, milia or fish, mattrima or sex, and mudra or symbols."

"I saw an Aghor sadhu down by the river using his fingers and thumbs to form symbols that were related to some sort of sexual practice."

"Yes, they use their fingers to form eight signs for *linga* and several more for *yoni*." Swamiji places his thumbs together and then his fingers, stretching them to form an approximation of a vagina, a sign I had seen Shankar make. "This is one of the signs they use," he says.

"What do they mean?"

Swamiji shrugs. "That's their secret, and you must become a sadhu to know."

"Why did you leave?"

"As I grew older I saw clearly and turned away from the evil inherent in the practice of Aghor. The Aghor sadhus follow Tantra's left-hand path, the way of the impure, and their mantras give them demonic powers. They worship devils. They live at the cremation ghats and converse at midnight with demons and witches. We Tantrics who follow the path of purity, the right-hand path, keep our distance from the Aghoris, and you should be very careful if you go near them. The sadhus may try to mesmerize you, to get you in their clutches. Don't eat or drink anything they give you. Their holiest men eat human flesh, and this is not easy to practice. How they are able to overcome repulsion and do it without going mad is because of Aghor mantras they keep secret. They eat abhorrent things like human flesh, their own feces, and drink their own urine, and they say that when you have no reaction to the taste then you are near enlightenment. As Tantrics, they have no wife or children."

He sits beside me and touches my arm. "You must also be careful if they take you to their most important ashram, Kinaram, which is not far from here. They don't cremate their leaders when they die because they have become samadhis, the most pure of the pure, the holiest of the holy, the enlightened ones, and so they bury them in the grounds of Kinaram. That makes it a very dangerous place for those who don't believe in their teachings."

"I'm going to Kinaram tonight."

"Then, I'll pray that the Aghoris don't harm you, and that you return safely."

He escorts me outside and says goodbye. I agree to come back in three days.

☠

My trip to Kinaram, the Aghoris' mother temple, had been arranged by Ron Barrett. He had put me in touch with Bantu Pandey, an important Aghor devotee who lives in Benares. He has business interests in the United States, and has agreed to take me to Kinaram. Just before seven that night, Pandey, a tall middle-aged Indian on a motor scooter, arrives outside my hotel. I climb onto the back and we roar off, weaving smoothly through the inevitable traffic jam, to Kinaram. The gate of the temple, which the Aghoris also call Krim Kund, is festooned with huge concrete skulls, twenty times life-size. As we remove our shoes at the gate, Pandey tells me that he runs a business that brings young American devotees from California to Benares to visit Kinaram and other holy places.

I had seen the Californian ashram's website, which urged its followers to practice the Aghor teachings of "simple living and being a good human being." There were plenty of mushy platitudes, such as "the pursuit of good character and good behavior turns a human into a God." However, there was nothing about the cannibal holy men revered by followers of Aghor in India, and their preference for living by cremation grounds and smearing themselves with ash from the bodies.

"Do your Californian followers know that Aghor holy men eat human flesh?' I ask.

Pandey ignores the question, but offers his own experience in striving to be a good Aghori. "Ron Barrett and I decided one time when he was here that we'd seek out a corpse and attempt to meditate while sitting on it, like our holy men. We searched along the riverbank for six days until we found a corpse that had been washed up. It had been swept down from somewhere upriver. We tried to get close, to sit on it, but the smell was too terrible, and we gave up. That's why I admire our sadhus because they have the mental strength to overcome human sensitivities to such things."

"Do you feel the same about them eating human flesh?" Once again Pandey ignores my question.

Inside the temple grounds, more giant concrete skulls are dotted amid the winding footpaths, tall trees and mouldy buildings. Pandey points to the grave mounds of the sect's ruling gurus, now samadhis, or men who attained enlightenment and bypassed the karmic cycle. He explains that the body of such a person cannot be cremated, but must be buried. "There are many Aghor samadhis buried here," he says. The present-day Aghor leader, Baba Siddharth Gautam Ram, is not at the ashram. "He comes and goes, and our people believe he appears out of a hole in the ground here at the ashram." Pandey yet again ignores my question, whether he believes the guru has this power.

Edged against the temple is a pool, about four times Olympic size, filled with water coated with green slime. "The pool's water is sacred, and if you immerse yourself in it, the water will cure all your ills. Many people come here on Sunday to be cured, and barren women who go into the pool soon after become pregnant."

Nevertheless a highly revered Aghor leader, Baba Bhagwan Ram, suffering from diabetes, journeyed to the United States to find a cure before dying of complications from the disease in 1992. Pandey remains mute when I ask why the guru didn't immerse himself in the pool's sacred healing powers.

We come next to the temple's main hall. It has no walls, the roof is held up by columns, and it boasts a large painting of the sixteenth-century founder of the Aghor sect, Baba Kinaram, garlanded with skulls. At the other end of the hall is a life-size statue of the Guru encased in glass. A pair of pretty young girls sit cross-legged on the bare floor seeking guidance from a shaven-headed, bare-chested sadhu clad in a white sarong who looks like a malevolent Uncle Fester of the Addams Family. The holy man sits by a small hearth piled with ash from the cremation ground at Manikarnika. He thumbs ashes on their foreheads, and then whispers a response, causing them to bend close to him to hear.

A wedding party arrives to seek a blessing, the young groom in an ill-fitting suit and a crown made from stiffened silver paper, his bride in a red sari embroidered with gold thread. Her face is hidden by cloth in the traditional way, and she clings to the back of her mother, who maneuvers her about the temple. Five bridesmaids in orange saris trail behind them. About fifty devotees join the wedding party in a small courtyard for *aarti*, the evening prayers. Six musicians blow conch shells, beat hand drums and clang bells, as a young priest offers clouds of incense from a brazier to the samadhis' burial mounds and to the barred inner sanctum of the founder, where the Aghoris claim a sacred flame he lit still burns.

Pandey points to the bride and bridegroom, who are clearly from low-income low-caste families. "That's one reason why Aghor is growing in popularity in India, because people can escape their caste and be treated equally for the first time in their lives. We have one hundred and fifty ashrams all over India, as well as the ashram in California, and plan more."

As we leave, Pandey also warns me that there are many Aghor sadhu imposters in Benares who perform for the pilgrims. "But Anil Ram Baba at Harishchandra cremation ghat is a true Aghor sadhu," he says. I don't tell him that I already know Baba and plan to meet him again. My questioning has irritated him, and I do not want him to warn Baba to stay away from me.

☠

The next morning my guide Tiwari takes me to meet His Majesty of the undertakers, the Dom Rajah, king of the Doms all over India. His home is in the old city, overlooking the Ganges. We walk through a maze of alleyways, passing by hundreds of tiny shops offering pilgrims the essentials of Hindu piety; holy pictures of Shiva with triton and hand drum; his monstrous wife Kali holding a severed head and wearing a garland of skulls; and India's most prolific playboy, blue-skinned Krishna playing a flute in a verdant glade to captivate his hordes of adoring milk maids.

Ever since I made my first visit to India decades ago my favorite Indian god has been the elephant-headed Ganesh, the god of knowledge, son of Shiva and Parvati, and his portrait is clearly in demand. Marigold garlands, bowls of incense, trays of vermilion powder and jugs of coconut milk neatly set out in hole-in-the wall shops are popular with the pilgrims. Many walls in the old city are decorated with murals of the gods, especially Hanuman and Ganesh, but most are faded and mouldy, attacked by the summer's intense heat, and then by daily gales of rain when the monsoon arrives. Again and again I hear the chant of burial parties coming along the lanes, and step aside as they pass by bearing their charges to eternity.

The Dom Rajah, Jagneesh Chowdhury, lives in a concrete house high above the river, guarded by a pair of life-size tiger statues on a balcony overlooking the Ganges. He is immensely wealthy, but lives as simply as a farmer. A pair of well-fed cows stand in a small outer courtyard, to give the appearance of Hindu sanctity, but also to provide his family, as he tells me, with much tastier unprocessed milk. The Dom king is a stocky man with a huge belly, and his mouth seems filled with blood. He is chewing *paan*, or betel nut and a sprinkling of lime wrapped in a small leaf: it is India's favorite narcotic. Chewed, the result is a scarlet liquid that pleasantly numbs the senses.

Jagneesh leads me into a small room dominated by pictures of Hindu gods and goddesses that line the walls, and sits on a bed. Women scurry in and out of the room, but he ignores them. He holds in his arms his little son, the Doms' crown prince, a sulky child whose eyes are heavily ringed with black kohl. It gives him a scary look, like a demonic child in a horror movie.

His speech slurred by the *paan*, the Dom king begins by telling me the creation story of his people. Ages ago, Shiva and Parvati came to Benares, and Parvati went to bathe in a pond. A Brahmin named Kalu made the mistake of entering the pool at the same time, embarrassing Parvati. Shiva, angered by his transgression, punished the Brahmin by demoting him to a low caste. Ashamed, Kalu apologized and asked what kind of work he could do now that he had been reduced to such low status. Shiva told him that from that moment on he and his descendants should live at Manikarnika ghat and tend the sacred flame for cremations, collecting money for its use. "That's how we became Doms," he tells me.

Jagneesh pauses to spit a stream of scarlet liquid onto the floor. Moments later a servant edges crab-like through the door, his body bent close to the floor, and cleans up the mess. Even Untouchables have their own Untouchables. "Hindus must be cremated using the sacred flame, and we Doms have had sole possession of it for thousands of years," Jagneesh tells me. Everyone—"maharajahs, prime ministers and billionaires all come to us in the end." The poor pay a rupee if that's all they can part with, but the rich will pay one hundred thousand rupees or more. But "it's the same flame for all."

We move out on to the wide balcony with its views of the slushy Ganges and the bluff with its many ghats, temples and palaces. The religious cacophony down below is like the murmur of the wind. Jagneesh, still holding his son, leads me to one of the tiger statues. A century ago the Maharajah of Benares picked a fight with Jagneesh's grandfather, who was the Dom Rajah at the time. "My grandfather put one of the tiger statues on the balcony overlooking the Ganges to show people his power. The Maharajah ordered him to take it down, claiming that only the royal family had that privilege. My grandfather refused. 'Your ancestors were nothing thousands of years ago when we Doms were given the sacred flame by Shiva,' he replied. The maharajah took my grandfather to court and lost. To mock the ruler, my grandfather placed a second tiger statue here."

☠

Tiwari and I leave the Dom king and plunge back into the maze, heading for the Golden Temple. We stop at a small stall to drink tea laced with cardamom. For the past few days all India has been obsessed with the coming marriage of Bollywood superstar Aishwariya Rai to Abishek, son of Amitabh Bhachhan, the country's favorite movie star for three decades. I had met Bhachhan on a Bollywood set a few years earlier and witnessed major movie actors treat him with dumbstruck awe, as if he were a demigod. Many rural Indians consider him just that. The wedding is daily front-page news, and one TV channel is covering it twenty-four hours every day.

Aishwariya and Abishek had just visited a Benares temple with their families, to be blessed by the priests, spotlighting for the nation the importance of the city's spiritual prominence. I ask Tiwari about a report I had read, that Aishwariya had to be married to a tree by Brahmin priests before she could go ahead with the marriage to Abishek.

Tiwari shrugs. "It's true. No matter how educated, no matter how attuned we are becoming to Western ways, we Indians are still bound by traditions and customs that first formed thousands of years ago." He promises to introduce me to an astrologer who can explain why Aishwariya had little choice but to marry the tree first.

We head once more into the alleyways in search of Benares's eternal spirit, but encounter its modern tragedy. A narrow lane leads deep into the medieval city by the ghats, its entrance guarded by a dozen paramilitary policemen armed with assault rifles. As we delve deeper into the maze, we encounter more than one hundred policemen and soldiers in camouflage uniform protecting important temples, including the shrine of Kali the terrible.

"It's most necessary," says Tiwari, his face tightening. "Last year Muslims bombed the monkey god temple, and the railway station, killing twenty people."

Since then security has been tight in the old city. Although Benares is Hinduism's most sacred city, it is still 25 percent Muslim, and it sits in Uttar Pradesh, India's biggest state, which has witnessed many bloody clashes between Muslims and Hindus. The old city itself is a provocation. When Aurangzeb, the Mughal Emperor, conquered Benares in 1669, he flattened almost all the buildings along the Ganges to rebuild the town for Muslims. "He tried to rename the city Muhammadabad, but the name didn't stick," says Tiwari. For Hindus his most grievous crime was to demolish the city's most important shrine, the Vishwanath Temple, dedicated to Shiva. The Muslim Emperor used the stones to erect a mosque on the spot. "Aurangzeb was brandishing Muslim power over Hindus," says Tiwari. "It was a wasted effort."

When the British deposed the Mughals in the nineteenth century, Hindus rebuilt Vishwanath, which became known as the Golden Temple after a Maharajah paid to have it coated with eighteen hundred pounds of twenty-four-carat gold. At today's prices that would be worth about twenty-three million dollars.

Devotees flock to the temple to witness the five daily *aartis* during which priests pray before one of India's holiest objects, the most sacred of the many *linga* of Shiva in Benares, an upright phallic stone. They bathe it with Ganges holy water, and then anoint it with milk, sandalwood paste and a heady mixture of honey and *bhang*. An Indian army major gripping a pistol commands soldiers clad in camouflage searching the long line of devotees wanting to enter the temple. Other troops grip assault rifles and scan the throng. "We fear another terrorist attack on the holy shrines," the major tells me.

Tiwari, perhaps moved by being so close to the godhead, reveals that he is active in the militant Hindu political group, the Bharatiya Janata Party (BJP), which came close to provoking an all-out war between India's Muslims and Hindus when the party held national power from 1999 until 2004. The deputy prime minister then was Lal Krishna Advani, a dangerous rabble-rouser. He is suspected of earlier urging a Hindu mob to demolish a sixteenth-century mosque in Ayodhaya in Uttar Pradesh in 1992, because they believed it had been built over a Hindu temple almost five centuries ago. There is no time limit on the murderous tension between Indian's Hindus and Muslims.

"The Muslims killed many sadhus in revenge," Tiwari tells me. "Anyone with a prayer mark on the forehead became prey." Riots erupted across India with more than three thousand people, mostly Muslims, slaughtered. Advani was India's federal opposition leader at the time, and commentators accused him of exploiting the Hindu-Muslim rift with provocative speeches that turned his supporters into bloodthirsty mobs.

After the incineration of fifty Hindu train passengers in 2002, an act linked to the temple slaughter as belated revenge, hundreds of Hindu mobs attacked Muslims in planned pogroms across the country. Edward Luce, in his book *In Spite of the Gods*, published in 2006, described how "mobs gathered around and raped the [Muslim] women, they poured kerosene down their throats and the throats of their children, and threw lighted matches at them. Hundreds stood by and cheered these gruesome incinerations, which symbolised revenge for the burning of the train passengers."

At the mobs' forefront was the BJP's youth wing, the Rashtriya Swayamsevak Sangh (RSS), which trains members in aggressive self-defence. "I was an RSS member when I was young because I wanted to defend Hindus from the Muslims," Tiwari tells me. "They're a violent people, and we're fighting for our faith."

In two weeks, Uttar Pradesh is holding a state election, and the Indian political world's megastars, including the Gandhis, India's Kennedys, are descending on the holy city for election rallies. Eight of India's fourteen prime ministers came from Uttar Pradesh, and the state is a testing ground for high office. The city is slathered with political posters, and cars carrying loudspeakers blare out political slogans day and night.

"Mr. Advani will be at a rally tomorrow night, and you should go," Tiwari says.

CHAPTER 11

Boning up on the BJP rally I am astonished to find that criminals dominate the Indian parliamentary system and its august federal and state chambers. It is as if the Mafia controlled the US political system and openly stacked the legislatures with their candidates. The magazine *India Today* reported in a pre-election story that "over 80 percent of the state's current MLAs"—members of state legislative assemblies—"have criminal cases against them. Some of them even stand from jail and win."

Uttar Pradesh is not unusual. When I contact Anil Mishra (a common surname in Uttar Pradesh), a reporter with the popular TV channel Sahara, he tells me surveys have shown that more than 50 percent of India's politicians, state and federal, have criminal backgrounds, with many charged and convicted for murder, rape, extortion, kidnapping and threatening to kill anyone brave enough to stand against them. "Here in UP, Satya Pradesh has forty-five criminal charges pending against him including murder and rape. Muktar Ansa, a Mafia don, is alleged to have murdered a politician, but he's contesting this election from his jail cell. People are impressed by their acts of criminality because it sets the perpetrators apart as tough guys, like Bollywood heroes."

The jailed candidates use a loophole that allows them to stand for political office if they do not have a conviction. The Indian court system is notoriously slow, and a case can go on for a decade or more before a verdict is reached. One enterprising gangland don, Afzal Ansari, a ruling party candidate in this election who is on trial for murder, called a rally from the jail using his cell phone. An aide placed a phone next to a loudspeaker at the rally, allowing the incarcerated politician to address the crowd.

As darkness falls across Benares the following evening I join more than a thousand BJP supporters surging through the narrow streets like a river in flood, going to the Advani rally. Indians are an exuberant people, and their zealotry can best be seen when their political passions swell and they take to the streets. The BJP supporters, clad in saffron shirts, the Hindu holy color, beat drums and blow conch shells. "Jindabad BJP!" they shout "Long live the BJP!" The throng is so high-spirited that even bearers carrying bodies to the ghats step aside as we pass. Only the holy cows refuse to budge.

BANG! The sudden blast stuns me and I drop to the road, fearing a terrorist attack on this fundamentalist Hindu rally. A smell of gunpowder soaks the air. But I am the only one on the ground. People smile down indulgently at the mad foreigner. "It was a very big firecracker," a sadhu brandishing a triton tells me. "We use them to announce our presence. Jindabad BJP!"

Thrice more the air is torn by the explosive's bang as we near the *maidan*, an open field where more than ten thousand BJP supporters are already seated. Looming over the *maidan* is the town hall, a stately maroon Gothic building put up by the colonial British, and policemen line its high walls. A bystander tells me they are snipers. Several hundred more policemen ring the ground.

Benares is a focus for Hindu nationalism, and the rally could provoke an attack by Muslim extremists or foreshadow a Hindu rampage through the Muslim quarters. Before the rally, according to the *Times of India*, the BJP issued a DVD and an ad campaign that called Muslim Indians unpatriotic and branded them as terrorists. So, the police are taking no chances.

BJP worthies sit cross-legged on a stand as if at a prayer meeting. As we wait for Advani, rabble-rousers fire up the crowd by screaming anti-Muslim slogans through bullhorns. The crowd stirs and the conch shells blare as Advani arrives in a procession of cars and bodyguards. From the stand he offers his supporters a *namaste*, the traditional self-effacing greeting with hands joined together, but even I can tell that his humility is a ritual pose.

India's most dangerous politician looks like a government clerk in a neat black waistcoat, tight white pants, and a tunic that ends at his knees, but up close he crackles with charisma and his eyes have a dark ruthless glint.

Senior BJP politicians on the stand rush forward to drape massive garlands of flowers around his neck and offer fawning *namastes*.

The firebrand comes straight to the point, demanding that the mosque at Ayodhya be pulled down and a Hindu temple erected there as "a matter of cultural nationalism." Advani knows that if that happens communal riots will break out across India once more, and the country will be awash with the blood of innocent Hindus and Muslims.

In days to come I will see Italian-born Sonia Gandhi, the ruling Congress Party's titular head, and her thirty-seven-year-old son, Rahul, the family political heir-apparent, at open-air rallies in Benares. Neither comes close to rivalling the crowd appeal of Advani. Rahul seems a nice young man, enduring on-the-job training in the family business, a hazardous occupation. His grandmother, Indira, was assassinated, and so was his father, Rajiv, both when they were prime minister. Tiwari told me that many Hindus believe the Gandhis to be soft on Muslims, and Rahul must fear that if he is ever elected to the highest office, each time he steps out in public a bomber or a gunman could be there, aiming to kill him.

None of their gilded promises of a better life for all seems to affect the relentless grind of poverty. I read in *The Times of India* a prediction by demographers that urban poverty rates are growing twice as quickly as total urban populations, and the urban poor will comprise more than six hundred million people by 2025. Advani's baiting of Muslims does nothing to ease the lives of the urban Hindu poor, but it is clearly effective as a way of swelling his party's popularity.

The next morning, turning away from the madding parade of Indian politicians and their empty slogans, I ask Tiwari to take me to a famed astrologer. We're in a city that probably has more astrologers per head of population than any other place on Earth, and I am still intrigued as to why Bollywood's Aishwariya Rai had to marry a tree before she could tie the knot with her true love.

Dr R.M.P. Mishra is the principal of a Sanskrit college and a revered astrologer, consulted by politicians, movie stars, businessmen and professionals from across India. He ushers me into his study and smiles at my bafflement. "Aishwariya has the planet Mars in a central position in her horoscope, and this means her first marriage will inevitably be troubled and end in divorce. All Mars people suffer from this. So, to sidestep the problem, following Hindu tradition, she was married to a tree. But no ordinary tree, it was a pipul. It represents the Hindu trinity, with Shiva the leaves, Rama the roots and Vishnu the trunk. Once her marriage to the tree ended, she was free to marry again without fear of trouble because of her astrological chart."

I ask Dr Mishra to cast my horoscope, more in fun than for any need to see into my future, and am amused when he looks at my right hand palm and expresses the same surprise and delight as Shankar the Aghor sadhu. He traces with a finger the triton and hand-drum, unmistakable signs to him that I am a special favorite of Shiva. "You can achieve anything you want because Lord Shiva will always be at your side," he says. "He has drawn a protective circle around you and anyone who creates problems in your life, Lord Shiva will knock them down."

Then, in a ninety-minute reading, after learning of the time and date of my birth, he astonishes me by knowing many pivotal events in my life, too personal to mention here, giving the time, place and significance of each incident. Either astrology is a true science or he has an eerily accurate power of ESP. That prompts me to ask, a touch fearfully, about my future. "You'll live to ninety-one because the moon and sun in your chart give you much energy and keep you free from disease. Your whole life you'll be active. If you decide to do something, no one can stop you. You have one child now, but you will have a second."

The spell is broken. I express disbelief that I will make it to ninety-one, not sure that I want to; and scoff at his last prediction because I have no plans for another child. "It's not your choice," he smiles. "The gods have ordained it, and you can't erase it from your chart."

☠

A few days later I return to Harishchandra ghat at midnight bearing a bottle of whiskey. A body is being burned by the river, and once more the red glow guides me down the darkened steps to where Anil Ram Baba sits by his hearth. He accepts my offering with a smile and wastes no time in pulling out his *kapalik*, the top half of a human skull, and filling it with the whiskey. He downs the liquor, refills the skull and again gulps down the whiskey. I tell him about the many warnings I have received to be beware of him, and he shakes his head with a grim smile. "You're safe with me," he says. "You're the first foreigner who's come to the ghat at midnight to talk with me, and your interest pleases me."

He marks my forehead once more with ash from the cremation pyre, but this time does not offer to put it on my tongue. His Dom friend is not here, gone home to his wife and children, and this seems to relax Baba. He begins to chant a mantra, but when he finishes does not explain what he used it for.

He begins to tell me his story. Baba is forty-two, and has been an Aghor sadhu for more than twenty years. He lives by the ghat, and never leaves, save for an occasional trip to his guru's ashram downriver. He depends on the generosity of devotees to eat and drink, and as a sadhu cannot marry. The whole world, he tells me, is his family, and he prays every day for their welfare.

"Do you sleep with women?"

He nods. "Women followers offer themselves to me, and sometimes I accept. It's a test of the control I have over my body and mind because it's forbidden for me to let free my semen. I must never finish the sex act even though we can have sex for many hours. I know special mantras that allow me to do this. That pleases many women because they know they're not in any danger of becoming pregnant when they have sex with me."

"What about prostitutes?"

"I've no need for them because there are more than enough women wanting to have sex with me. I've also practiced a sex act that few sadhus ever attempt. About three years ago, I obeyed my guru's command and sought out the dead body of a woman to use for sex. When females have been killed by a snake, or smallpox, or who've died when pregnant, they are not cremated but have a rope with a heavy stone tied around them and are then taken by boat out into the river and dropped into it. Soon after my guru commanded me, one night I was sitting here and saw a woman's body rise to the surface of the river. It usually takes two or three days after they've been dropped there. I swam out to the body while chanting the mantra my master had given me for this moment. Despite the look of the body, she was a whiteish color, a little bit blue, this mantra caused me to become erect and the effect lasted for about an hour and a half. I gripped the woman's body and got on top of her and had sex with her for all that time."

Because I had become used to stories about eating human flesh, somehow this tale made me queasier than from anything I had heard about cannibals. "Why do such a thing? Is it because it's one of the most monstrous things a person can imagine?"

"That's why I did it, because it's not monstrous, that judgment is just an illusion in people's minds. By overcoming this I come closer to enlightenment."

"Have you eaten human flesh?"

"Many times here at the cremation ground. We call it divine food, *maha prasad*. I can go for weeks or months, and then suddenly hunger for it. When I see that some flesh has not been consumed by the sacred flame I go down to the pyre and take it. The family do not object. But before I eat the flesh I chant special mantras that my master taught me. Without them a person eating human flesh would go crazy. It tastes very good, especially the brain."

There it is again. I think of Bailom the Korowai and the lustful gleam in his eye when he told me that of all the human parts he enjoyed eating the brain best.

The next day I join Swamiji, my Tantra teacher, for the final lesson. He steps up to the whiteboard and outlines the general principles that have illuminated the first three lessons. I listen with an open but skeptical mind because I continue to believe it to be mostly a hodgepodge of religio-philosophical babble, cleverly formulated and never able to be tested for proof.

"Now, let me tell you about empowering snake power, *kundalini shakti*," he says with a smile. I edge forward on the sofa.

In a book entitled *Tantra, The Path of Ecstasy*, published in 1998, the author, Georg Feuerstein, wrote that stirring this snake power requires "contracting the sphincter muscle, and by applying the throat lock while holding the breath which causes the *prana* and *apana* to mix and 'combust,' thereby driving the life force upward into the central channel." That's not something I've heard Sting mention when he speaks of practicing Tantric sex.

"Before sex the sadhu makes the *yoni* drink," Swamiji says, and presses on without an explanation of this intriguing practice. Drink what? Water? Diet Coke? "Then, with the proper mantra, the Tantric can perform the sexual act for hours at a time, but he must never come to the climax."

"Doesn't that cut out the intense culmination of sexual pleasure?"

Swamiji ignores my aside. "It's important with serpent power that the woman gets exhausted during the sex, but not the sadhu."

"Why not?"

"A man's semen is life itself and has an immense effect on the stability and power of his mind. So, his semen must be preserved."

"That doesn't sound like much fun."

"Sex is not meant for fun when you are a sadhu," he chastises. "It's to practice control of his semen. To do this, a sadhu must learn to draw up to a gallon of water up through his penis and hold it there for several hours." Feuerstein described the same technique.

"Do you practice this?"

"Of course, it's the way of a Tantric who wishes to become enlightened."

I decide at this moment that I have no wish to become enlightened, though I am intrigued to find out more about Tantric sex. Swamiji, innately modest or believing that I had only just a beginner's knowledge of Tantra, and so was not ready for the full description yet, keeps his explanation to just those few sentences. But the Tantra book is in some places explicit in its descriptions of its bizarre sexual practices.

"There is nothing glamorous about Tantric sex even when practiced in a group," Feuerstein wrote. "Ideally, all participants are spiritual practitioners." He tells of a boy who was initiated into the sexual secrets of Tantra by a woman he calls the mysterious "Lady in Saffron." In contrast to that denial of anything glamorous, the description of what followed throbs with allure. "Together they went to an abandoned temple where the [boy] lit a fire, threw incense into it, and then became deeply absorbed in meditation. Sitting next to her, he too closed his eyes and drifted off. Suddenly he felt her gentle touch and when he looked at her, he found to his utter astonishment that she was now completely naked, lying prone with her legs in the lotus position, and flower petals scattered over and around her genitals and with her pubic hair and other parts of the body besmeared with ashes and dabs of red and black color.

"The lady told him that he was the living flame, time eternal, the sun,

Brahman, while she was a corpse, time-bound, the sky and the lotus. Then she asked him to recite Sanskrit verses, and soon he lost all sense of her presence and even his own being. 'Something was happening to the mound of my penis. A vibration, thrilling, hot deep throb hammered beat after beat. The more it came in waves, the more I was pushing out my spinal base . . . a strange feeling of completeness, fulfilment, and ecstasy settled on my nerves.'"

Ahh, now I know why Sting and all those other rockers have with great diligence studied and practiced Tantric sex. The hour is fast coming to an end, and I remind Swamiji that he promised to teach me a mantra at this final session that will grant me my most fervent ambitions in life.

"I have not forgotten." He chants solemnly: "*Gun Ganapati Namah.*"

It doesn't sound very exciting, but I repeat it.

"The mantra means Salutation to Ganapati, another name for Ganesh, the god of knowledge."

My expectation drops with a thump. I was hoping to be given a mantra far more earth-shaking. "Say it seven hundred thousand times over twenty-one days, and you'll be granted everything you ever wanted."

Seven hundred thousand times in twenty-one days? That is thirty three thousand three hundred and thirty three times a day. Impossible. "It's too much, I can't do that," I tell Swamiji. "I've got to sleep and eat and work."

He looks at me with the sympathy of a master recognizing the frailty of a student. "Then do it one thousand times a day, and it will still be effective."

"That will take almost two years to complete the task. Does it take effect before I finish, let's say proportional to the number of times I chant the mantra?"

He shakes his head. "Only when you've finished the full seven hundred thousand times will you be granted the immense power it brings."

The chance is lost. As we walk to the door, Swamiji touches my arm. "Have you been with the Aghor sadhu at night at the cremation ghat?"

"Yes, I've visited Anil Ram Baba several times at midnight at Harishchandra."

"And he's not harmed you?"

"No, he's been very friendly and helpful."

Swamiji nods. "Then you truly are a favorite of Lord Shiva. He has been shielding you from the Aghori's evil designs on you."

☠

I visit Baba again the next day. He is sitting outside his hut, watching a body burn down by the river. Smoke spirals into a blue sky, and stoic relatives huddle to the side.

Baba looks as though he has been praying for most of the night, or perhaps he had got hold of another bottle of whiskey. His eyes are bloodshot and his face has the pallor of exhaustion. "I'm going to give you one of our holiest mantras," he says. "Chant it every day and think of me." He leads me down to the river where a boat is waiting. "I don't want passersby to hear it," he murmurs.

Dipping against the current, we row out to the middle of the river, beyond the many boats packed with pilgrims. Sunbeams speckle a path across the sacred water, and we are alone as we head for the far riverbank, the din from the bells, flutes and conch shells softening and then drifting away on the breeze. The many temples and shrines spilling down the ghats on the far bank shimmer like a mirage from a time far back in the past.

Baba halts the boat in the shallows on the far side of Ganges, a sandbar jutting from the shore. The riverbank on this side is bare of buildings. The boat bobs gently up and down for there is scant current over here. "I wanted to get away from the ghats because too many people there would want to listen to what I'm going to tell you," he says softly. "Now, write down the chant." He begins at a fast pace, and I have to ask for him to slow down. Not knowing Sanskrit, I can only render the mantra phonetically.

Tatom Guru Puram
Guru Sururam
Ruham Suham
Om Kring Kring
Hoong Hoong
Samsani Kali Ge
Kring Kring Kring
Hoong Hoong
Hring Hirin Suha.

And so the chant goes for about ten minutes, words that have no meaning for me, words that Baba says that he cannot translate. "Chant it every day," he tells me, "and you'll never be sick. Kali-Ma will protect you. She'll give you health and happiness for the rest of your life."

I am leaving in the late afternoon and back at Harishchandra cremation ghat Baba and I hold hands in farewell. He has shown me friendship when others warned that he would harm me, and perhaps lure me into Aghor cannibalism. They were wrong. Spurning the chance to tempt me into becoming a devotee, he has tried his best to help me understand the Aghoris' rituals.

Baba was not raised to eat human flesh, like Bailom, Kili-Kili and the other Korowai I journeyed with through the New Guinea jungle, and so I cannot forgive him his cannibalism. Bailom and Kili-Kili have known no other way from the time of their first thoughts. In the treehouse, killing and eating a *khakhua* is such a normal part of life that the participants in cannibal feasts not only do not believe they are committing an evil act but believe that they are doing good, convinced that their act destroys forever an evil presence that has come among them.

At one level they are no different from us. The United States, throughout its history, has put to death those it regarded as criminals, just like the Korowai do with the *khakhua*. America has also sent off to war soldiers with the intent to kill enemies, just like the Korowai warriors are obliged to kill the *khakhua*. Of course the Korowai then take ritual and legal murder to another level by eating the bodies of the executed.

Baba's cannibalism is different because he had free will, deciding to become an Aghor sadhu when he had just left his teens. The nature of what he would do in pursuit of holiness was recognized by most of those around him as evil, and in a pernicious twist of Aghor logic he became convinced that this evilness was just a human illusion. There could be no evil in eating human flesh, he believes, and that by breaking the taboo he comes closer to enlightenment. And, yet, as Baba and I say farewell, perhaps never to meet again, at this moment I do not judge him harshly. Perhaps this reflects my decades-long struggle on journeys into the remotest places on Earth to understand customs and behavior so utterly different to mine.

(Writing this months afterward, the emotion of being with Baba has long evaporated, and I am now harsher in my judgment of his behavior. Eating human flesh, unless you have no prospect of other food and are starving to death, is an evil act, justifying the taboos placed upon it throughout much of human history, and no amount of religious mumbo jumbo can sanctify it.)

I cannot escape the Aghoris in Benares because on the way to the airport, later that afternoon, the taxi passes by Kinaram, their mother temple, and as I glance at the giant skulls hung over the entrance they seem to be attempting to mesmerize me, compelling me to return to this sancta sanctorum of saintly cannibals. I turn away, free now from their dangerous allure, happy that I am heading for home.

Tonga's Warrior Cannibals

CHAPTER 12

My quest moves to the South Pacific island nation of Tonga, known as the land of the world's fattest people. What draws me are tales that Tonga was until recently the home of the Pacific's fiercest warriors, giants who feasted on the bodies of their slain enemies. They cooked them in *umu*, or huge underground ovens.

Tonga is several hours' flying time into the Pacific from my home in Sydney, to the northeast of Fiji, but most flights stop over in Auckland, New Zealand. It is at Gate 3 of the international airport there that I begin to understand that I am on my way to a most unusual place.

The flight indicator at Auckland shows that the jet to Tonga is leaving in just over an hour's time. I am carried along by a throng of passengers bustling toward the departure gates, bound for Los Angeles, Tokyo or London. Others swarm about the duty-free shops, like ants around piles of sugar, buying high-priced liquor and perfumes. After three decades on the road, this is home to me, nothing exotic. But then I turn a corner and find the Gate 3 waiting room crowded with nut-brown giants. I had heard that Tongans were very large, but it is another thing to see them close up. Men, women and children cram into airport seats that give up trying to contain their massive bodies. Arms, legs and buttocks overflow the armrests.

I sit one seat down from a tall beefy teenager who must weigh at least 280 pounds. Wrapped around his massive torso is a densely woven straw mat that covers him from chest to ankle, held up by a thick band of twine. He looks as if he has been gift-wrapped for a cannibal feast. For a moment

I think he might be a little crazy, but then notice that almost all the Tongans are clad in straw-mat wraparounds. I try to imagine the stir it would cause if I arrived at the airport with a chunk of lounge carpet tied around my waist.

"Gud ivining," he says in an ear-grating New Zealand accent.

The giant's name is Mapa and he explains that he was born in Auckland, but in blood and heart is and always will be a Tongan. "We nivir stop being Tongans, no mitter where we luv," he says.

Mapa lives (luvs) in Mangere, an Auckland suburb known as Little Polynesia, where the Tongan diaspora is large and thriving. About a hundred thousand people live in Tonga itself, with at least that many resident abroad, mostly in Hawaii, San Francisco, Los Angeles, Auckland and Sydney. Family ties are strong, and most of the passengers are going to Tonga to attend weddings, funerals or baptisms.

"My cusun und hus fumulee luv un Sydney? Feleti Moala. Ya know hum?"

As I explain to Mapa that Sydney has four million people and that, at most, I know just a few hundred, a pair of stewardesses arrive dressed in neat blood-red uniforms, the national color of Tonga. Not for Tongans the slim pert dream girl of Asian airlines. These girls flaunt the broad beams, bulky shoulders and sturdy limbs of champion hammer throwers.

Mapa reaches across and pokes me in the ribs. It feels like he has struck me with a war club. "Pritty gud lucking cheeks," he smiles.

I nod, not sure if he means chicks or cheeks, facial or otherwise, but try to see beauty the way the Tongans do. Basil Thomson, prime minister of Tonga in the late nineteenth century, observed in 1902 that female beauty there was valued more by the girl's substantial heft than any hour-glass figure. "The perfect woman must be fat—that is most imperative—and her neck must be short; she must have no waist, and if nature has cursed her with that defect she must disguise it with draperies. Her bust, hips, and thighs must be colossal. The woman who possesses all these perfections will be esteemed chief-like and elegant, and her nose will not matter, though, if she have that organ flat to the face, she will be painting the lily."

Judging by the display of beauty at Gate 3, nothing seems to have changed. Shakespeare's Julius Caesar worried that the lean and hungry are dangerous and he would have welcomed Tongans to Rome. In contrast to the nervous clamor or apprehensive silence of most of the other passengers at the airport, the Tongans at Gate 3 laugh, joke, smile and rib each other nonstop. It was a similar geniality that prompted Captain James Cook to call Tonga the Friendly Islands. On a visit in 1777, he penned in his journal, "the Tongan women are the merriest creatures I ever met with, and will keep chattering by one's side without the least invitation, provided one seems pleased with them."

The smiling, laughing, chattering passengers at Gate 3 seem the direct descendants of the women Cook encountered. And they look attractive in a regal way. As Tongans arrive at the gate, they walk with such grace and stateliness that, given their great height and massive bulk, they have the aristocratic look of matriarchal elephants. Nothing, I suspect, would hurry Tongans, not even an official racing through the airport shouting that the place was on fire. Once settled in their seats, they seem as immovable as monuments.

Boarding is slow as each Tongan carries two or three bulky gift parcels, and they block the aisle as they struggle to cram them into the overhead lockers in the 737. As they settle in their seats, jowl by jowl, buttock by buttock, beefy arm pressed against beefy arm, I wonder how these huge people can fasten the seat belts around their waists. Are we going to take off with most of the passengers unbuckled? What will happen at take-off if the pilot has to jam on the brakes in an emergency? Will I be in danger of being crushed by tumbling giants?

The stewardesses have a well-practiced answer, moving down the aisle before departure, handing out extenders giving each hefty passenger another half a yard of belt and buckle to handle waists that measure 50 inches or more. The extenders must be standard issue on any flight to or from Tonga.

Thankfully, I have been given an aisle seat. The husband and wife in the two seats next to me are encased in bulky straw mats and, together, must weigh at least 500 pounds. They both read from Bibles as we take off.

☠
⤬

Tongan history traces back to the tenth century (about the time of Alfred the Great in England). Until the middle of the nineteenth century the warlords, or *Tu'i*, were regarded as immortals, descended from the Polynesian gods. At the snap of a finger, they could have a flunky club to death anyone for the slightest insult.

Europeans first arrived in Tonga in 1643, when the Dutch explorer Abel Tasman sailed his ship through their islands. Many more European ships followed, and from them, stories of Tongan cannibalism began to slip out to the outside world. In 1789 when William Bligh, captain of the HMS *Bounty,* and his loyal crew were set adrift in the waters near the island of Tofua, just north of Tonga's main island, by the mutinous Fletcher Christian and his men, Bligh was afraid to take his men ashore. It was only after a few days and a dire need for water and food that Bligh finally overcame his fear of the Tongans' reputation for cannibalism and accepted an invitation to a local chieftain's feast.

The Englishmen had not been there long when Bligh sensed they were about to end up cooked and quartered on the feasting mat. He yanked out his cutlass and shouted an immediate retreat. He ran with his men back to the rowboat as scores of warriors charged after them. With Bligh and most of the crew safely in the rowboat, quartermaster John Norton went back to shore to release the anchor. The Tongans swarmed him and bashed his brains out with rocks. There is no proof, but he was almost certainly cooked and eaten by the warriors who killed him, following Tongan custom. Thus, Norton became the first victim of Fletcher Christian's mutiny.

The far-off Tongans gained further notoriety in England a few decades later when a young sailor, William Mariner, published a book describing how he spent several years with them. In 1806, his ship, a 500-ton privateer named the *Port au Prince* and manned by ninety-six sailors, put in at Lifuka, not far from Tofua, to careen the vessel and mend leaks in the hull. Mariner was apprehensive. A year earlier, the ship's crew had plundered the town of Hilo on the coast of Chile, wreaking havoc in the local church, stealing its consecrated vessels. Some months later, while the crew was being entertained by the governor of a port called Tola, Mariner described the theft to the governor's sixteen-year-old daughter. In great agitation, she "lifted up her hands and eyes, uttered some expressions in Spanish," and warned him that because of this "heinous sacrilege" he would never again see his mother and father, the ship would never return to England and the crew would be killed "as a just judgement from God."

Her prediction largely came true at Lifuka when more than three hundred warriors swarmed over the *Port au Prince,* clubbing to death most of the crew. Mariner, just fifteen, did not back down when confronted by "a short squab naked figure, about fifty years of age, seated with a seaman's jacket soaked in blood thrown over one shoulder; on the other rested his iron-wood club, splattered with blood and brains." Mariner was saved because the king, Finau, had recently lost a son in a battle, and Mariner's bravery under attack and his noble appearance stirred his admiration.

Mariner lived in Tonga for four years, and published the first of several editions of his memoir of his time with the Tongans in 1817: *An Account of the Natives of Tonga Islands* gives a rare eyewitness account of the murder and eating of captured enemy warriors after a battle. Mariner's detailed accounts of Tongan customs proved remarkably accurate, and that gives credibility to his description of a cannibal feast. "Some of the prisoners were soon dispatched," he wrote. "Their flesh was cut up into small portions, washed with seawater, wrapped in plantain leaves and roasted under hot stones. Two or three of them were also baked whole, the same as a pig."

Mariner followed this with a lengthy explanation of how he witnessed the cooking of the bodies, observing that after each victim's belly was sliced open, "the whole side of the carcass was next filled with hot stones, each wrapped in bread-fruit leaves. The carcass was then laid, with the belly downward, in a hole in the ground lined with hot stones. A few other branches were then laid across the back of the carcass, and plenty of banana-leaves strewn, or rather heaped over the whole upon which a mound of earth was raised so that no steam could escape. By these means, the carcass could very well be cooked in about half an hour."

Following this detailed recipe for preparing a human body for feasting, Mariner is almost casual in his reporting of other cannibalism among the Tongans. In the aftermath of a victorious battle where he fought at the side of his protector, the king, Mariner describes how the bodies of sixty slain enemy were laid out for a time in front of huts dedicated to various gods. Most were then taken away, but "nine or ten were conveyed to the water side and there disposed of in different ways.

"Two or three were hung up on a tree; a couple were burnt; three were cut open for motives of curiosity, to see whether their insides were sound and entire, and to practice surgical operations upon; and lastly two or three were cut up to be cooked and eaten, of which about forty men partook." Later in the same passage, however, Mariner relates that, "after their inhuman repast, most persons who knew it, particularly women, avoided [the cannibal warriors], saying, 'Iá-whé moe ky-tanga-ta,' away! you are a man-eater."

Mariner claims he spurned the offer of human meat, but he remarked that "the smell of it when cooked is extremely delicious." On his return home he settled into the life of an upper-middle class gentleman in London, improbably, given his Tongan war exploits, working as a stockbroker.

Almost a century later the Tongans were still killing and eating enemy warriors. In 1904, Sir Edgar Collins Boehm, an English explorer, published a book, *The Persian Gulf and South Seas* which described his nautical roaming among several Pacific kingdoms. In it he included a graphic illustration of Tonga, drawn from life, entitled "Cannibals dragging prisoners to the ovens." Two more illustrations showed Tongan "cannibals slaughtering their victims."

The Tongans weren't the only Pacific Islanders to lust after human flesh.

The people on the island of Tanna in the nation of Vanuatu in the South Pacific were infamous for being among the most violent and enthusiastic cannibals in the South Seas. The practice only ended two or three generations ago. This warrior from a tribe still living by choice in the Stone Age does a war dance. His recent ancestors would have been cannibals.

The neighboring Fijians were enthusiastic cannibals according to their tribal historians. (In fact, Mariner's book makes the claim that Tongan cannibalism imitated the Fijian practice.) So were the Marquesans, a bellicose people who lived and made war in a sprinkling of ten volcanic islands about eight hundred and fifty miles northeast of Tahiti. A.P. Rice in *The American Antiquarian*, published in 1910, wrote that the Marquesan islanders considered it "a great triumph to eat the body of a dead man. They treated their captives with great cruelty. They broke their legs to prevent them from attempting to escape before being eaten, but kept them alive so that they could brood over their impending fate."

Rice added that "when the hour had come for them to be prepared for the feast, they were spitted on long poles that entered between their legs and emerged from their mouths, and dragged thus at the stern of the war canoes to the place where the feast was to be held."

On most international flights, late at night you are lucky to get a biscuit and a tumbler of coffee in economy, but on a flight to or from Tonga, a full meal is always served, even at 2 am. It requires dedicated eating several times a day to maintain a Tongan's bulk. My reading has made it impossible to eat. The stewardess looks at me with tender sympathy as the roly-poly man seated next to me eagerly offers to eat my meal as well as his own.

The Tongans were never colonized, the only South Pacific nation to claim this distinction. The ferocity of the native warriors frightened off any challenge. Despite remaining free of colonial domination, the nation did not remain free of Western influence. Missionaries of several Christian creeds landed there, converting some of the inhabitants. Tongan history then shifted radically in 1875, when a Christian warlord, Taufa'ahau, already the most powerful chief in the realm, proclaimed himself King George of Tonga and introduced a constitution largely drafted by his friend, an English missionary. He was an admirer of British royalty, and swiftly turned the country into a South Pacific miniature of the British monarchy with similar titles and honors, and an elaborate court dress.

The new king cleverly reknotted Tonga's traditional Polynesian society by transforming its most powerful chieftains into hereditary nobles. Those thirty-three landed aristocrats occupy the second rank of a feudal three-class society that, even in the twenty-first century, is segregated by blood. At the feet of the royals and nobles squat about a hundred thousand "peasants" forbidden by law to marry anyone from the king's extended family. A noble who marries a commoner is stripped of his title. That the imposition of such a medieval European order of society was so successful must have had something to do with the lack of colonists. Had Tonga ever been under the British thumb, their people might have eventually rebelled against the outsiders and cast off their courtly ways.

King George's Tongan title was Tu'i Taufa'ahau and he married Princess Halaevalu Mata'aho. The names roll off my tongue like music. The great beauty of the Tongan language—soft, lilting, melodious, the speech of poets and musicians—makes an interesting contrast to the warlike people who speak it, gleefully recounting their epic battles and the cannibal feasts that followed.

Now, those cannibal feasts are slipping back into legend. Almost all Tongans are Christians, mostly fundamentalist Methodists, thanks to King George, who banned worship of the ancient Polynesian gods. The Tongans

The king's only daughter, Pilolevu, at the celebration of her birthday. Tongan society is strictly hierarchical and even high-born ladies must sit at the feet of the royals.

are now the most devout Christians on Earth. Each Sunday, the entire nation shuts down to observe the Sabbath. Fervent prayer services run from dawn until well after dusk. But for the other six days of the week the Tongans feast and love with an intensity that is pure Polynesian.

☠

Just after 3 am, the 737 sets down at a small tin-roofed airport where sea birds perch in the rafters. "Welcome to Fua'amotu airport in the kingdom of Tonga!" the captain announces. The airport nudges the ocean on Tongatapu, the country's most important island for the past six hundred years.

I emerge from the plane on the tarmac into a steam oven, the humid heat even in the early morning drawing beads of sweat all over my face. Burly, wide-hipped immigration officials examine our documents with the Polynesians' unhurried grace. Surrounded by so much corpulence, the passenger in front of me, a young woman, stands out. She is Tongan, but she is also tall, slim and shapely, unlike any other Tongan woman I had seen so far. On the way to the baggage collection area, her firm round hips, held captive by a red silk mini skirt, sway and bounce as she walks.

As we wait for the luggage to arrive, the girl smiles at me, and I return the compliment. "Hello!" she says in a sweet voice. "Good morning," I murmur.

The girl has finer facial bones than the other Tongans on the flight, the edges highlighted by bronze makeup, and her rouged lips are delicately shaped. Her eyelids are traced with mascara, and the hint of a duty-free perfume pleasures the air around her. I find just one thing puzzling. Her hands seem too large for the rest of her, as if a football player's massive mitts have been grafted onto the slick body of a New York model.

A cocky young man strides up as I emerge from the arrival hall. "G'day mate!" he says. "I'm Tavita. It's Tongan for David. Want a taxi?"

The accent is perfect Strine, the Australian dialect that often sounds like double-dutch to outsiders. My antennae quiver. He is the hunter and I am the prey. I accept his offer, intrigued by his accent, and his Aussie backyard barbecue gear—baggy shorts, bargain-basement T-shirt and flip-flops.

"I went to Sydney for my sister's wedding when I was seventeen," he explains. "I worked there on a garbage truck for six years until the immigration department caught me and sent me home."

"You know the International Dateline Hotel?" I ask.

He nods. "Sure! It's a dump, but it's the best hotel in town."

Suddenly, the arrival area is split by girlish screams. The beautiful girl at the baggage collection has just emerged, and is surrounded by other beauties in skimpy dresses that flaunt their long legs, slim hips, and that supposed curse to Tongan womanhood, curvy waists.

Tavita sees me staring at the girls and shakes his head. "You better stay away from them, they're fuck-a-ladies," he says.

"What did you call them?" I whisper. This is the world's strictest Christian country, and I glance around to see if any sharp-eared police officers are near. The coast is clear.

"They're fuck-a-ladies. You know, men who think they are women."

When he sees that I am still puzzled, and taken aback by his blunt language, a sly grin slips onto Tavita's face. "Sorry, you foreigners always get the words mixed up. It's not what you think. It's spelt f-a-k-a l-e-i-t-i. Faka in Tongan means to act like, and leiti is the way we spell lady. So, the *fakaleitis* are men but they act as if they are women. They dress like them, act like them, even make love like them. Get it?"

Before, no. I could have sworn the pretty young miss was really a girl. But now, yes. That explains her less than delicate hands. As we pass by the giggling *fakaleitis*, they still look and sound to me like beautiful young girls, but most have large hands, and up close I see that many have a bulging adam's apple.

"Bye, bye, darling, see you soon!" the girl from the baggage hall calls out.

Tavita hurries me past her. "Plenty of foreign men come to Tonga to sleep with the *fakaleitis*, but they're real trouble, more bitchy and unfaithful than real women. You'd better stay away from them."

As we walk to the car park, I assure him that I am not at all tempted, though I would like to meet the girl and her friends to find out about their lives and how they became *fakaleitis*. "I won't have anything to do with the buggers," he growls.

His aging white Toyota is a rust bucket and would be pitched onto the scrap heap in most countries. "I named it after my wife, Mele, because she's not the prettiest girl in town, but she's all mine," he says as he pedals the accelerator, prompting a groan from the engine. "I wanted my sister to find me an Aussie girl I could marry and settle with in Sydney, but on a trip back here I met Mele, fell in love, and that was the end of that dream."

Tavita is about five feet seven, and weighs about a hundred and fifty pounds. By Tongan standards he is a pygmy and I wonder if Mele is bigger. "She's taller and fatter," he says happily. "My mother is Samoan, and they're usually not as big as Tongans."

I settle back in the taxi for the forty-five-minute journey to Nuku'alofa, the capital of Tonga. Its name means "abode of love." Within sight of the airport, the shell of a half-built hotel lies abandoned in a field surrounded by waist-high grass and framed by coconut trees. "That's our famous airport hotel," Tavita says with a grin. "It was financed by a Chinese, Dr Sam Wong, who had more cash than good sense. Fancy building an airport hotel on an island where the only town has about twenty thousand people. The King supported him even though we all knew it was a foolish idea."

"I'd heard that Tongans revere the king, and believe he can do no wrong."

Tavita greets this with a snort. "The old people still love him, they still believe in the old ways. They say, 'Koe fonua o'o Tupou mo Ho'eiki.' That means, 'The country, property and people all belong to the king and the nobles.' But most of the young people know that's bullshit. We're sick of the king and his family because they've been ripping off Tongans for decades. They've taken hundreds of millions of dollars out of the country. You remember Suharto in Indonesia? Our royal family is the Tongan version."

"The Indonesians overthrew Suharto," I say. "Any chance of that happening here?"

"Fat chance," he answers with a sly grin, enjoying his own joke.

The elderly taxi forges on, grunting and wheezing, threatening to fall apart as it bounces over a rutted road hemmed in by palm trees silhouetted against a purple sky. Jungle flows into jungle, broken here and there by villages and more churches than I want to count. In some the lights are on, despite the early hour, and I see through open doors worshippers on their knees before the altars.

On the outskirts of Nuku'alofa, a car pulls by us, crammed with young men and women gripping bottles of rum. Sunday is still twenty hours away. "Hello!" the driver shouts, waving a bottle out the window. "Welcome to Tonga!"

"It's just like Captain Cook experienced when he came here," I remark as the car roars ahead. "He found Tongans to be more friendly to strangers than any other people he'd encountered."

Cook came to Tonga with two ships, *Resolution* and *Discovery*, on his third Pacific voyage in 1777. He wrote in his log book: "This group I have named the Friendly Archipelago as a lasting friendship seems to subsist among the Inhabitants and their courtesy to strangers intitles [sic] them to that Name."

"You believe that?" Tavita asks with a derisive laugh. "We Tongans are good liars. The king tricked Cook into thinking he was welcome. Sure, he gave him plenty of feasts, but he was playing for time while he decided how to kill him. The king and his nobles planned to grab all the treasures Cook's ships were carrying, and then eat him and his men. But they argued over whether to attack by day or at night, and left it too late. Cook sailed away never knowing about the plot. The silly bugger."

Cook did not witness any cannibalism on Tonga during his three visits, but on his first Pacific voyage, after sailing on from Tonga to New Zealand in October 1773, at Charlotte Sound he described in his log book how some of his officers went ashore and made a gruesome discovery. "They saw the head and bowels of a youth who had lately been killed, lying on the beach and the heart stuck on a forked stick which was fixed to the head of one of the largest canoes. One of the gentlemen bought the head and brought it on board, where a piece of the flesh was broiled and eaten by one of the [Maori] natives before all of the officers and most of the men. I was ashore at this time, but soon after returning on board was informed of the above circumstances."

He was "filled with horror" at the sight of the severed head, but "curiosity got the better of my indignation . . . and being desirous of becoming an eye witness of a fact which many doubted, I ordered a piece of the flesh to be broiled and brought to the quarter-deck, where one of these cannibals ate it with surprising avidity."

Cook was far ahead of his time in attempting to understand and respect the often (to him) bizarre and sometimes gruesome habits of the indigenous people he came across in his Pacific voyages. He wrote of Maori cannibalism:

> Few consider what a savage man in his original state and even after
> he is in some degree civilised; the New Zealanders are certainly in a
> state of civilization, their behaviour to us has been Manly and Mild,
> shewing always a readiness to oblige us; they have some arts among

them which they execute with great judgement and unwearied patience; they are far less addicted to thieving than the other Islanders and I believe strictly honest among them-selves. This custom of eating their enemies slain in battle (for I firmly believe they eat the flesh of no others) has undoubtedly been handed down to them from the earliest times and we know that it is not an easy matter to break a nation of its ancient customs let them be ever so inhuman and savage, especially if that nation is void of all religious principles as I believe the New Zealanders in general are and like them without any settled form of government; as they become more united they will of concequence have fewer Enemies and become more civilised and then and not till then this custom may be forgot.

In Tonga Cook and his men themselves missed by a very whisker of fate ending up the same way as the Maori victims whose body parts they saw being eaten. Lucky bugger indeed, for a time. Six years later, at Hawaii, on Valentine's Day, warriors swarmed over Cook in the shallows at Kealakekua Bay, stabbing and clubbing him to death. The warriors carried away his body, which they butchered.

Lieutenant James King, one of the ship's officers, wrote poignantly of the aftermath: "About 8 o'clock, it being very dark, a canoe was heard paddling towards the ship. There were two [Hawaiian] persons in the canoe, and when they came on board they threw themselves at our feet and appeared exceedingly frightened. After lamenting with abundance of tears and loss of 'Orono'—as the natives called Captain Cook, one of them told us that he had brought us part of the body.

"He then presented to us a small bundle wrapped in cloth, which he had brought under his arm. It is impossible to describe the horror which seized us, finding in it a piece of human flesh about nine or ten pounds weight. This, he said was all that remained of the body of 'Orono.'"

The Hawaiian claimed that the rest of Cook's body had been cut into pieces and burnt—all but the head and bones, which were in the possession of Terreoboo the king. The chunk of flesh, wrote King, that they "were looking at had been allotted to Kaoo, the chief of the priests, to be made use of in some religious ceremony. He said he had brought it as proof of his innocence."

Vanessa Collingridge, in her book *Captain Cook*, published in 2003, relates that six days after Cook's murder, a procession of priests and villagers carried to the beach a package containing most of the captain's remains and gave it to Charles Clerke, who had taken command of the expedition after Cook's death. Inside it he found "the scalp, the skull minus the jawbone, all the long bones, thighs, legs and arms. There were no ribs or spine or feet, but his hands . . . still had the flesh attached, which had been slashed and stuffed with salt for preserving."

On the following morning, the Hawaiians delivered to Clerke another package containing Cook's jawbone and feet. All the body parts were placed in a coffin and weighed down, and toward sunset buried at sea.

The Hawaiians may have been concealing a dark secret. They were known cannibals, and they believed that to eat the flesh of enemies imparted into the eater the eaten's bravery and *mana*, or spiritual charisma and power. We will never know but, after sailing unharmed for more than a decade among man-eaters, it is possible that Cook ended up as one of their victims.

CHAPTER 13

T avita is right. The International Dateline Hotel is a dump, and I am an expert on hotels that are dumps.

It is just after 4 am and the reception lobby looks promising. It has a simple Polynesian elegance that hints of a Somerset Maugham novel, the kind of place where you might run into Marlon Brando or Errol Flynn in the old days, on the prowl on a remote South Pacific island. Open and spacious, it has a stone floor, a teak check-in desk, large framed pictures of the gargantuan king and his bulky queen. It looks onto a lawn, coconut trees, beds of tropical flowers and a small swimming pool.

A coffee-skinned Tongan girl clad in a blue sarong hands me the check-in form. Even at this impossible hour her smile is neon-bright. Nestled against her right ear is a red hibiscus, in Hollywood lore a Polynesian signal of a girl's availability.

"If you put the hibiscus behind the right ear, does that mean you're taken, or you're available?" I ask with a smile.

She measures me with a confident look, and returns the smile. "Want to find out?"

My cheeks turning crimson and Captain Cook's admiration for Tongan women shared, I walk up two sets of stairs to the room. It is tiny and shabby. The ancient air conditioner's wiry entrails have been torn from its belly, and it perches in a hole hacked into the cracked concrete wall. The window facing the ocean does not shut properly, leaving a gap through which the cold air can escape. But that is not a problem because the air conditioner does not work.

The bedroom, hot as an oven, is about three paces wide, three long. The paint peels from the wall as if it has caught some tropical disease, and cigarette burns pockmark the small table. The bathroom, with shower and toilet, is dark, grimy, claustrophobic and the lair of giant mosquitoes that whine like fighter planes as they hurtle toward their targets, my bare arms and legs. The Tongan Visitors' Bureau has awarded the hotel three stars. They must have forgotten the minus sign.

☠

Eager to get out on the streets, I set the alarm for 7 am. When it rings, the sun has yet to gather its oppressive heat, and the waterfront outside the hotel is thronged with Tongans arriving by bus and van for the Saturday market at Vuna Road. The scent of the frangipani trees lining the street mingles with the Pacific's briny tang and the musky odor of coconut oil the Tongans rub into their hair and on their bodies.

Large smiling Tongans sit cross-legged in the straw-top open stalls offering in-your-face hip-hop T-shirts from the US, portable radios, boxes of soap powder, car tires as wide as their sellers' girths, plunkety-plunk Islander ukulele music on cassettes and stacked tins of baby food.

The smell leads me to the fish stands where the night's catch is on display. At one stall, a fisherman with bloodshot eyes squats by his catch, a tiger shark. The marine man-eater is about eight feet long and lies on a bed of palm leaves, its black eyes as hard in death as they must have been in life. A stick props open its cavernous mouth studded with rows of sharp milky-colored teeth. The shark's satin skin shimmers under the rising sun's rays.

As on the plane, I am surrounded by giants. Adults and children, they are all gargantuan, with most women weighing at least two hundred pounds, and most of it is blubber. Their massive thighs threaten to burst out of their leotards or sarongs, and their bull-like chests strain against sheet-sized T-shirts or floral dresses.

The men often melt to fat as they grow older, but when young seem carved from a block of granite. Across the road I meet twenty-six-year-old Tuula Vana, who says Tongan men are traditionally tough. "In the old days our men sailed outriggers to nearby Fiji to go fighting for the fun of it."

Tuula is six feet eight inches tall with muscles on his muscles. If I saw him coming at me with a Tongan war club, I would drop dead with fright.

"Did the warriors eat the fallen enemy after a fight?"

"Of course they did," Tuula says with a grin. "It was good tasty meat, so why waste it?"

In these fierce battles, fallen Tongan warriors were just as likely to end up as meat in a Fijian underground oven, a *lovo*, just like the *'umu*. The Fijians called their victims *bokola*, or cannibal meat, a word that becomes a grievous insult when flung at another person. When a revered warrior or chief was killed, chunks of his flesh were dispatched to chieftains across Fiji. In 1867, when the British missionary Thomas Baker was murdered on the island of Viti Levu, each important chief across the island was presented with a piece of his flesh to eat.

A sandalwood trader who plied Fijian waters between 1808 and 1809, Captain Richard Siddins, wrote about the butchering he witnessed of a chief killed in battle. "The chief's hands were cut off at the wrists, his feet at the ankles, his legs at the knee, and his things at the middle. The head was then cut off very low toward the breast and placed on some hot ashes."

The cannibals cooked the meat in the *lovo*. "Once the head was cut off, then they cut off the right hand and the left foot, right elbow and left knee. And vice versa until all the limbs were cut off. The guts and vitals were also taken out and cleaned for cooking. They cut the flesh through the ribs and right to the spine, which was then broken—halving the body. The cleansing and preparation of the body took about two hours. The flesh was then wrapped in plantain leaves and placed in the oven."

There is much similar evidence to show that the Polynesians and their Pacific neighbors the Melanesians were enthusiastic cannibals. American anthropologist Peggy Reeves Sanday made an extensive study of cannibalism. In *Divine Hunger: Cannibalism as a Cultural System,* published in 1986, she described one of the most compelling eyewitness reports she had encountered, an 1879 report by a native of the Cook Islands who became one of the first Polynesian missionaries. His travel log and letters both describe the consumption of human flesh. "One particularly lurid but descriptive example comes from a report of a war that broke out in New Caledonia soon after his arrival there as a missionary:

I followed and watched the battle and saw women taking part in it.
They did so in order to carry off the dead. When people were killed
the men tossed the bodies back and the women fetched and carried
them. They chopped the bodies up and divided them . . . When the
battle was over they all returned home together, the women in front
and the men behind. The womenfolk carried the flesh on their backs;
the coconut-leaf baskets were full up and the blood oozed over their
backs and trickled down their legs. It was a horrible sight to behold.
When they reached their homes the earth ovens were lit at each house
and they ate the slain. Great was their delight, for they were eating
well that day. This was the nature of the food. The fat was yellow and
the flesh was dark. It was difficult to separate the flesh from the fat. It
was rather like the flesh of sheep.

I looked particularly at our household's share; the flesh was dark
like sea-cucumber, the fat was yellow like beef fat and it smelt like
cooked birds, like pigeon or chicken. The share of the chief was the
right hand and the right foot. Part of the chief's portion was brought
for me, as for the priest, but I returned it. The people were unable to
eat it all; the legs and the arms were only consumed, the body itself
was left. That was the way of cannibalism in New Caledonia.

She writes further that man-eating even sent Fijian tribal cannibals into
a frenzy of sex after bodies of enemies taken from the battlefield had their
sexual organs, both male and female, sliced from their bodies and hung in
sacred trees to rot. "As the flesh cooked, the warriors and women danced
the lewd death dance. As the women danced naked, they compared their
genitals with those of the stripped carcasses of the enemy warriors, praising
the sexual prowess of the killers, and sexually insulting the naked enemy
bodies. Feeling mounted during the cooking and feasting until the tension
was broken in a frenzied sexual orgy."

☠

Spying chunks of pink fleshy pig meat on sale at the Nuku'alofa market, I
pass on breakfast, and after two cups of coffee to chase away my weariness
from lack of sleep, stroll down Nuku'alofa's main street. The downtown's
few paved streets are lined by trade stores, dimly lit and dusty wooden
and ferro-concrete shacks selling fishing tackle, three-week-old Auckland

papers, 1950s-style cotton floral dresses and shorts and other South Pacific treasures. Cars even more dilapidated than Tavita's desperate taxi crawl along the main street looking is if they will fall apart at the next bump. Telegraph poles tip to the side, threatening to topple to the dusty ground.

That night the air conditioner in the room still refuses to work. I change rooms, but the rusty relic cemented into the wall roars and shakes, ineffectively battling the steamy night air. Sweat bathes my body, it is hard to breathe and, although it is almost midnight, I succumb to an urge to go to a bar. At least there I am not likely to die of heat exhaustion.

Tavita comes to the rescue. "I'll take you to the best bar in Tonga: the Blue Pacific."

"If it's anything like the best hotel in Tonga, you needn't bother."

He flashes a pearly smile. "I told you the Dateline was a dump."

The taxi's doors, windshield and undercarriage rattle with each bump in the rutted road as we drive to the far end of the island. The Blue Pacific perches at the edge of a lagoon where Captain Cook first stepped ashore. The nightclub is strung with flashing lights, and coconut trees tower over it, silhouetted against the midnight sky. A muscled bouncer bursting out of his black suit pockets the six *pa'anga* entry fee. "Welcome bro', plenty good lookin' girls in there tonight," he says with a leery wink, waving me inside.

A Tongan song set to a disco beat blasts from the speakers. About thirty tall young men and women are dancing with graceful rhythm. The youngsters, clad in jeans and sarongs, weave beautiful patterns in the air with their beefy arms, and slip their sturdy hips to and fro in time to the thumping beat.

When the music stops, two men and three girls grab chairs and invite me to drink with them, ordering rum and coke, and vodka. The girls, especially, are friendly, drawing me into the conversation, treating me from the first moment as if we had been friends for years. They honor an ancient tradition. Tongan women were even more welcoming when they flocked to see the first Europeans to land here in 1643, the Dutchman Abel Tasman and his crew on a journey of discovery from Java through the South Seas.

Tasman's barber-surgeon Henrik Haalbos pioneered the fashion of South Seas journeyers tantalizing men back home with tales of its beautiful promiscuous girls. He wrote in his journal about two Tongan women who "both grasped me round the neck: each desired fleshly intercourse: whereupon they assailed each other with words. Other women felt the sailors shamelessly in the trouser-front and indicated clearly: that they wanted to have intercourse. Southlanders, what people."

I could not even draw a tickle. These modern Tongan girls prefer their rum and coke to a pasty-faced Australian. After a couple of vodkas I tell them about my recent journey to the Korowai cannibals in New Guinea. Their eyes widen at my gruesome tales of *khakhua* slaughtered by streams and carved up for eating. Only a buxom sloe-eyed girl with rounded lips and dark wavy hair secured in a bun is not impressed. "Just two generations ago my ancestors killed and ate our enemies, so your story is no big deal," she says scornfully.

She has no hibiscus behind her ear. Her name is Sosefina Halafihi, but she prefers Fina. "Sosefina sounds too soft," she says with a defiant look, as if she would bash me over the head with a war club if I disagreed.

Fina offers to meet again to help me understand Tongan culture. I agree, but after four days pass she has yet to contact me. Tongans, as true Polynesians, scorn schedules and deadlines, tuning their biorhythms to what they call "Tongan Time." The day before I had arranged to meet a new friend, a doctor, at two in the afternoon for coffee. He turned up at eight without apology, or even the sense that he was late. I held my tongue. Anyone demanding punctuality in Tonga is regarded as mentally impaired.

So, it does not surprise me when Fina fails to call. Four days could mean four days if you are lucky, but it could also mean a week or more. Tavita, in the meantime, has found me a feast, a favorite Tongan pastime. "It's our biggest obsession," he tells me. "We attend dozens of feasts every year, and you'd be astonished by the amount of food consumed. It will give you a sense of what the cannibal feasts were like."

☠

On a steamy afternoon, Tavita drives me along a palm-fringed seaside road to a wedding at Halaleva village. There I meet Tangoi Folaumoetui, a young woman working in the Prime Minister's office. Like all the women at the feast, Tangoi is clad in a *ta'ovala*, a formal straw mat that extends from her chest to her ankles, like those worn by the passengers on the plane. The tightly woven strands strain at her every move because Tangoi is very much on the superior side of two hundred pounds.

"I have to go to so many feasts that it's impossible to lose weight," she

A typical Tongan feast, this time a wedding with the bride in the pink blouse. Tongans are always going to feasts.

smiles while leading me to our seats, wobbling the trio of chins that flesh out her massive throat.

Shielded from the fierce sun by a canvas canopy, about a hundred guests sit elbow to elbow before long trestle tables that buckle under piles of food. For each quartet of guests there is an entire roast suckling pig surrounded by plates stacked with potato salad swimming in mayonnaise, greasy mutton flaps, plump cuts of chicken coated in oily batter and many more Tongan delicacies. Soft drinks wash down each of the dozen or more courses.

Tangoi passes a plate stacked with fat-streaked corned beef floating in a creamy coconut sauce. "Eat up, there's plenty more," she says, her buttery moon face beaming at the piles of food facing us. Eyeing the pig, she tears off a strip of the crackling skin and gobbles it down. In South Pacific folklore human flesh was known as "long pig." I suppress a shudder wondering if the way Tangoi attacks the pig meat resembled the way her recent ancestors enthusiastically consumed the bodies of their slain enemies.

The expression "long pig" was popularized in the West by Robert Louis Stevenson in his book *In the South Seas,* first published in London in 1900, which chronicled his extensive sea journeys through the South Pacific. In a chapter entitled "Long Pig—A Cannibal High Place," the eminent Scottish writer who had settled happily in Samoa in 1890, another cannibal stronghold, noted that, "cannibalism is traced from end to end of the Pacific, from the Marquesas to New Guinea, from New Zealand to Hawaii, here in the lively haunt of its exercise, there by scanty but significant survivals."

Stevenson seemed attuned to his homeland's squeamishness when he wrote, "Nothing more strongly arouses our disgust than cannibalism, nothing so surely unmortars a society; nothing, we might plausibly argue, will so harden and degrade the minds of those who practice it."

At first glance this was a Eurocentric view, uncharacteristically shallowminded in a man who ordinarily strived to understand the imperatives of the bellicose cultures he encountered in the South Pacific and their bloody rituals. There is a good chance that he would have read William Mariner's account of the cannibalism he witnessed in Tonga.

Stevenson was in the Marquesas when he wrote the words above. Even more curiously, he judged that the island cannibals he encountered "were not cruel; apart from this custom, they are a race of the most kindly; rightly speaking, to cut a man's flesh after he is dead is far less hateful than to op-

At the same feast. Middle-aged Tongans showing their fat.

press him while he lives; and even the victims of their appetite were gently used in life and suddenly and painfully despatched at last." He clearly did not meet the same cannibals in the Marquesas as *The American Antiquarian's* A. P. Rice.

At the wedding feast in Nuku'alofa, a *matapule*, a "talking chief," rises to praise with flowing oratory the dimpled bride's dazzling beauty and the stocky soldier groom's bravery and strength. Great orators are revered in Tonga and *matapule* follows *matapule*, the non-stop bellow of words drowning out most attempts at sustained conversation. That allows the guests to concentrate on the serious business of eating.

Tangoi keeps thrusting food at me. Mounds of ice cream and jelly. Giant taro tubers swamped with oil. Chunks of sweet potato floating in butter. Two hours into the feast, with my waist now dumpling shaped, I feel as if the overflow is about to burst out of my ears.

"I'm so full, I won't eat for another month."

"Don't be silly," she smiles. "This is just the start. Tomorrow, there's a big feast celebrating the first birthday of a friend's son. I'll take you there." Tangoi is true to her word and the next afternoon, after another mammoth feast that turns me green at the gills from over-eating, I seriously consider calling an ambulance to rush me to hospital. Half a bottle of Eno's fruit salts loosens the blockage.

Feasts have always played a vital role in tribal Tongan life, a social glue bonding friends, neighbors, even enemies. Whenever the chiefs put on a feast for Cook and his crew, the mats were spread with yams, breadfruit, pig, bananas, turtle, shellfish and fish. It sounds like the South Seas version of the Pritikin diet. You do not get fat on that kind of food, unless you devote yourself to a lifetime of overeating. But the men did not dare then, because a fat warrior was a stupid warrior and would be targeted on the battlefield. The commoner women toiled daily in the boiling sun, keeping their bodies free of flab.

Gluttony was reserved for the chieftains' wives who were pampered from childhood, fed colossal amounts of food, much like Peking ducks, so that they grew grossly fat. What better way for a chieftain to flaunt his wealth and power than to parade obese wives who never had to work his fields. Thomson's description of Tongan beauty in 1902, the perfect 10, had no neck and no waist, and would have resembled more a chieftain's wife than a woman who toiled at the clan plantation.

In Western society, rich women almost always prefer to be slim, and so we have deified bone-thin super models as the ideal beauty. Urged on by their "sisters" in the media, most other women work hard to keep down the curves. But in tribal Tonga, only the women from a high family could become very fat, and that became the ideal. The recent change to a cash economy has evened the scales, and, nowadays, any woman with enough money can stuff herself with food, at home and at feasts, until she looks like a chieftain's wife. Gluttony has become a nationwide vice. By her size alone, Tangoi, a woman from an upper-class family, would have made the perfect wife for a chieftain.

☠

And, of course, the perfectly fat woman nowdays desires to marry a perfectly fat chieftain. The most famed of Tonga's feasters, appropriately, is the king, eighty-seven-year-old Taufa'ahau Tupou IV. The six-feet-three-inch ruler was a champion athlete as a boy, but too fond of feasting, he became a champion eater. The 1976 *Guinness Book of World Records* listed him as the world's heftiest monarch, weighing a sumo-worthy 440 pounds. He is not shy about his girth. When visiting London, he rides about in his limousine boasting a license plate emblazoned "One Ton."

Taufa'ahau Tupou is one of the world's last monarchs with near absolute power, and, as I had already seen, he sits atop a rigid social order. This is clearly on view on Sunday in the Free Wesleyan church, "the king's church," near the palace. As I head there, I cross an empty main street on which all the shops are shuttered until midnight by the King's command because today is the Sabbath. The king has also banned all public activities that are not church-based.

A government publication aimed at tourists spells out the ban. "The Sabbath, from midnight Saturday until midnight Sunday, is a day of rest in Tonga. The law says it 'shall be kept holy and no person shall practice his trade or profession, or conduct any commercial undertakings on the Sabbath. Any agreement made or document witnessed on the Sabbath shall be counted null and void, of no legal effect.'"

In the side streets, bulky Tongan commoners move slowly toward the royal church. The shimmering air is so thick with humidity that they must push their way through. Wrapped from chest to toe in *ta'ovalas* that greatly restrict leg movement, they waddle like giant penguins up the church steps. Chattering noisily, they settle into the pews. At the front a brass band toting gleaming instruments clomps into the front rows by the altar rail. Behind them the choir, all giants, file in, creaking the knee rests with their great weight. About fifty strong, the men are clad in white coats, bow ties and wraparound lap-laps (cotton garments similar to sarongs) that end just below their chubby knees, while the women are modest in shiny white satin dresses that tickle their ankles.

A black Mercedes pulls up by the church to empty a noble and his wife onto the steps. Tall and fat are the traditional marks of the upper class in Tongan society, but this noble is a small, thin, neat man with a ferret's worried face. He is clad in a grey suit and, with a large pair of black sunglasses masking his eyes, looks like a clerk posing as a movie star.

"Many nobles wear dark sunglasses, even at night," Fina had warned. "We're not sure if they're ashamed to look at us because they're ripping off the country, or because they don't want to look unless they have to."

His highborn lady is impressively fat, handsomely jowled with ham-sized thighs. The aristocrats file into the church using their own side entrance, and join three other nobles and wives in elevated pews by the altar, peering down at the commoners from the godly side of the communion rail. Even in the house of the Lord, the nobles lord it over commoners. As I kneel in a pew, just behind the choir, it occurs to me that if Jesus came to church this morning, he would be with us, forbidden to enter the nobles' pews by the altar.

There is a hush as the king enters by his own entrance and, bent over double by his huge weight, climbs the steps to the elevated royal box overlooking the altar. He looks like a giant beetle. The queen, imperious, pale-skinned and large-nosed, follows, clutching a prayer book.

At the front of the church, the trumpets, trombones and tubas strike up a hymn, thrusting it forward with a thumping beat. The choir responds in full voice, battling the bandsmen to see who can make the most noise. But what glorious noise, as good as any Welsh choir. The sound thundering from the barrel chests of the huge Tongans shakes the pews. Woven among the harmonies of the baritones, tenors and sopranos is a high-pitched Polynesian keening that sharpens familiar hymns with an exciting edge.

Cook was the first outsider to notice and admire the Tongans' melodic talents. On his arrival in 1773, when he had his men play the bagpipes as a gesture of friendship, Tongan girls responded with "a musical and harmonious tune," surely a symphony of opposites.

The queen religiously mouths the hymn's words, but the king, impassive to the spectacle, pulls out a file from a briefcase and begins to read. He turns the pages, even when the pastor, built like a rugby forward, and with glowering eyes and shaggy mop of grey hair, ascends the pulpit. The priest glares at his parishioners for a few moments and then delivers a fiery sermon in Tongan, his deep powerful voice making a full frontal assault on the timber and mortar.

The king never once glances toward the pulpit. In mid-sermon, with a shrug, he closes the file and eases it back into the briefcase. Believing that God had granted him the divine right to rule Tonga, I expect that Taufa'ahau Tupou feels no need of guidance by mere men in the ways of the Creator.

When the service ends, he climbs into a king-size Cadillac and departs for his traditional Sunday ride along the main streets of his realm. It is an eerie spectacle. Seated in the back, his vision cloaked by dark sunglasses, the king looks straight ahead, his line of sight bouncing off the driver's neck as the vehicle crawls along Vuna Road. No Tongans line the streets to cheer him; no one even stops to look at the monarch, wave, or pay him respect as he drives past.

After morning service ends, Nuku'alofa is like a city of the dead. No person walks on the streets. No child cries out, even behind the closed doors. The port has been shut down since midnight, and the airport is silent, with no planes allowed to leave or arrive. Bathing at the many beaches is forbidden no matter how sultry the weather, fishing is taboo, and all sports are banned for the day. Even the birds perching on the trees and sagging electricity wires seem silenced by the king's command.

But at mid-afternoon, the ghostly silence is broken as church bells begin to peal triumphantly across the town. Tongan families, clad once more in *ta'ovalas*, waddle out of the bungalows and move ponderously toward their churches like herds of elephants heading for pasture.

☠

Once a Sunday is church enough for me. That afternoon I sit by the hotel pool reading a biography of the king's mother, Queen Salote, which is Tongan for Charlotte. She was one of the few Tongans ever to capture world attention. Much like the British Queen Elizabeth II, she was thrust at age eighteen onto the throne in 1918 by the early death of her father. Assuming that Salote Mafile'o Pilolevu was a guileless girl and easily led, the nobles married her to a bull-like noble named Viliami Tupoulahi Tungi who was expected to lead the tiny nation from behind the broad straw-mat skirt of his wife. Salote, the spawn of countless cannibal warlords, was made of sterner and smarter stuff, and began a sure-handed rule that only ended four decades later when she died suddenly of cancer.

In 1953, Salote traveled to London for the coronation of Elizabeth II. Tall, dusky and regal, the Tongan queen's Polynesian charm dazzled the British. She won their hearts when she insisted on riding in an open carriage through chilly London rain in the coronation procession, smiling and waving as if it were a sunny summer's day. Queen Elizabeth and all the other dignitaries rode in carriages with the hoods down.

Salote was splattered with rain, but, steeped in Tongan royal custom, she endured the discomfort to show humility and deference to the higher-ranked British Queen. Seated opposite Salote in the carriage was a shivering Malay sultan, his gold turban, tunic and pants soaked. "Cold. Get wet. Close roof?" he pleaded.

"I was naughty," Salote later confided to a senior British civil servant. "'No understand. No speak English,' I replied."

Noel Coward fixed the incident forever in cannibal legend when, during the procession, he was asked who was the tiny dignitary in the carriage with the massive Tongan queen. "Her lunch," he replied.

CHAPTER 14

Six days after my arrival, Fina finally calls with the name and phone number of the king's private secretary. He slots me in to meet Taufa'ahau Tupou the following day. Salote's son lives, as his mother did, by the sea in a nineteenth-century whiteboard palace, sharing the lush lawns and Norfolk Island pines with khaki-clad palace guards and the royal geese.

Prefabricated in New Zealand in 1867 and shipped to Tonga, the two-story palace, with its white gables, wide verandas, high steeple and gazebo looks like a superior Victorian guesthouse transplanted from Brighton's shores to the far side of the world. It was the dream home of its designer, a petit bourgeois English missionary in the South Seas who wanted to dazzle the locals in their grass huts with the splendor of European civilization.

Tongans are not so easily cowed these days. At the gate next morning, a soldier armed only with his own imposing bulk stops me. "Take off your sunglasses," he demands.

Rudeness begets rudeness. "Why?" I snarl back.

It is eleven in the morning, a bad time to be out in the sun in the tropics, and the sun stabs at my eyes. The palace is fifty yards across the lawn, and there seems no reason to suffer eye strain even for a moment. I had already bowed to the protocol demanded by the king's private secretary and donned a suit and tie for the first time in months as a sign of respect for the monarch.

"It's obvious," the soldier snarls, dropping the Tongan's friendly nature, unused to a visitor questioning his order.

"Not to me it isn't."

"I don't care what you think. Take the sunglasses off, or you can't go in."

A courtier, an aged balding man clad in a *ta'ovala* and long white-sleeved shirt, leads me, squint-eyed, across the lawn to the front veranda, up the steps, and along a corridor to the monarch's private study. It is a pleasant cosy room lined with bookshelves, the weatherboard walls adorned with holy pictures, the kind of retreat a missionary might favor to compose Sunday's sermon. One of the holy pictures shows a white-robed Jesus hovering over the gargantuan king with protective hands placed on his bulky shoulders.

I hear the king coming before I see him, moving ponderously through the palace, trembling the floorboards with every step. I stand as Taufa'ahau Tupou enters. He is clad in a grey tunic and *tupenu huluhulu*, the traditional skirt for males, and up close looks more like a statue of himself than a living man, immense, monolithic and detached. His dark impassive eyes barely register as he eases onto a seat.

The Tongan king, Taufa'ahau Tupou IV, in his study in his palace at Nuku'alofa.

Eighty-seven-year-old Taufa'ahau Tupou is on a diet. "I'm down to 133 kilograms [about 290 pounds], but I had to give up favorites, ice cream and Chinese meals," the monarch sighs in a deep voice that swells from his huge chest. I lean forward because the words are hard to distinguish. His face is so padded with flesh that the inside of his mouth thickens each sound so that it emerges slurred.

Nuku'alofa hums with rumors that the king is turning senile, but in a wide-ranging conversation he belies this with an impressive memory, skillfully analyzing the world's current geopolitical troubles. Six decades before he had earned a law degree at the University of Sydney and clearly retains a bright disciplined mind.

Having smoothed the way with conversation about the king's pet topic, I come to mine. "I've read that for centuries Tongans were cannibals, until just a few generations back. Is that true?"

The king remains silent, looking at me but not seeing me. He wears a watch on each wrist and the minutes tick by. His long majestic silences when meeting visitors are notorious; he has been know to sit for an hour or more during an audience with barely a word exchanged. The poor visitor sits on the edge of his seat, nervous and flustered, unsure whether to risk an act of *lèse majesté* by filling the edgy silence with his own voice, or mutely suffering until the king speaks.

Today he has been unusually voluble, the answers have sprung from his mouth, and so in gratitude I choose the second option and wait patiently for an answer.

"Of course those ancient Tongans were cannibals, it was part of our culture then," he finally says.

"Who did they eat?"

"Usually war captives, and it was only the warriors who ate human flesh, cooked in an *umu*."

"What about Captain Cook? I've been told that your ancestor, the high chief, planned to kill and eat Cook and his crew when they visited here on a peaceful mission."

Once again I am treated to a long silence with the king resting his massive jaw on his fleshy chest. His breathing comes in heaves of air, as if he has trouble lifting it from his throat and thrusting it into the ether. Minutes tick by, but the king remains immobile. Finally, he takes a deep breath and lets loose a deep-throated chuckle. I wait for the words to follow, but the king seems done with me. Moments later a flunky comes through the door with a glass of guava juice for me, a signal that the audience with His Majesty is at an end.

"Come to the gym this afternoon if you want to see me work out," the king murmurs in parting.

I have never seen a king pump iron, and even though it has nothing to do with my quest for Tonga's man-eaters, I accept the invitation. Three hours later, at the gym, Kala'uta Kupu, the instructor, snaps to attention as a large black car pulls up outside. The king emerges clad in baggy green shorts, white T-shirt, red socks and huge gym boots. The octogenarian is still a bull of a man, with enormous thighs and giant biceps, but he is doubled over by arthritis and hobbles silently into the gym gripping metal walking sticks. I am moved by how lonely he looks in public, his face set in stone.

The king settles into a leg-press machine and is immediately surrounded by the royal guard. One soldier sets the weights at 144 pounds; another cradles a small fan and directs it at the monarch as he begins to exercise. There is a soldier to count off each repetition; a soldier standing by to catch the king if he slips; and two more hover by a throne-sized chair to carry it to him in case the king needs to rest.

For forty-five minutes the king moves along the line of exercise machines with his praetorian guard. Then his physician takes his blood pressure. "A healthy 120/80," he murmurs.

☠

There is a message from Fina waiting when I arrive back at the hotel. "I'll call by at six tonight," it says.

At nine, the phone rings. "I'm downstairs." Even though we have only met the one time I would know that honey-soaked voice anywhere. Fina is clad in a yellow sarong that follows the curves of her hips. Although solidly built there is not an ounce of fat on her, and there is still no hibiscus behind her ear.

"Let's go celebrate your meeting with the king at the Blue Pacific. I've got a friend who wants to meet you. She says she knows you. Her name is Sabrina."

"I don't know any Sabrina."

The king working out in the gym at Tonga's capital, Nuku'alofa. The king held the world record for being the planet's heaviest monarch and had to work out constantly, even into his eighties, to keep reasonably healthy. At all gym sessions he is surrounded by the palace guard, ready to help him. One soldier directs a fan at the king as he exercises.

"She told me she flew from Auckland with the *palangi*, you. She's a *faka-leiti*."

"Why did you call me 'pawangee,' or something like that," I ask Fina as Tavita drives us to the Blue Pacific. "Is it a rude word?"

"No," she laughs. "It's *palangi*, our word for European. It comes from *papa* and *langi* meaning people who come from the sky. It was the only way we could explain the Europeans' big ships, strange weapons and fair skin when they first arrived about three hundred years ago. The name stuck."

"So, I'm a heavenly creature?"

"They were, but you look more like a devil," she says with a sly smile.

"Paaaaa'ulllllllllaaaaaa!"

The scream rattles my eardrums. It is the girl from the plane, now on the dance floor. She abandons her burly male partner and wiggles toward me. Her pink hot pants fit her like skin, her stilettos rise like skyscrapers. Most

Tongan girls have broad, buttery faces that melt into jowls and several chins once they have children and the accumulation of feasts mounts. But the *fakaleiti*'s face is like a wild cat's with angular, high, sharp cheek bones, and sloe eyes that grip as if you are prey.

"Hello Pa'ula, I'm Sabrina," she says in a voice unmistakably that of a man trying to be a woman, the vocal cords stretching under tension to raise the pitch.

"How do you know my name?"

"Fina told me," she says, fluttering her long eyelashes. "We've been friends for a long time. I lived at her house for a year, doing housework in exchange for a room."

We dance to an island love song strummed on ukuleles and sung in a falsetto. I feel like an SUV trying to do wheelies with a Ferrari. Clomp, clomp, clomp I go, thumping my feet to the lilting beat, as Sabrina flutters around me, slim arms dancing in the air, long legs moving the pert half moons of her bottom into a sensuous dance of their own. If she were not a man, with big hands and an adam's apple, Parisian designers would beg her to strut their catwalks. They still might if they ever see her.

Those beautiful eyes grip mine with a come-on that simmers with seduction. But she has a hibiscus perched behind her left ear—"taken," it says, and so I hope this is all pretend.

"Pa'ula dances divinely," Sabrina purrs as we join Fina, who is seated at a table with three more *fakaleitis*, all clad in short skirts and skimpy tank tops that reveal the upper part of their plucked flat chests. Tonga is one of the world's poorest countries, and expensive hormone treatment that prompts the growth of breasts in men—let alone a sex-change operation—is beyond the reach of the *fakaleitis*.

"Liar!" Fina laughs. "He looked like a hippo out of water."

"Ooh, you are nasty," Sabrina coos to me. "I like big-boned men."

"Don't you mean men with big bones?"

Sabrina slaps Fina lightly across the face. "No dirty talk here, my dear. I can see that Pa'ula is a decent man."

Tongans are hard drinkers, even the women and *fakaleitis*, and we toss back several rounds of rum and Coke, the drink of choice at the Blue Pacific. The talk is all gossip, mostly of loves won or lost or abandoned, but sharpened with jealousy and a catty lack of compassion for rivals.

"Whore," "Bitch," "Brute," "She-devil." The words, spat out, punctuate each tidbit. The *fakaleitis* contort their faces into cruel masks as they talk, but through it all, the smile lingers on Fina's lovely lips, though she trades gossip with gusto.

Sabrina, a little sozzled, loosens control over her voice so that it deepens and darkens as she leans against me and rubs her leg against mine. "Don't you think I'm sexy?"

"Very much so."

She drops her head so that she has to look up at me, even though she is at least five inches taller. "You want to take me home tonight?"

"Uhh, I have a very important meeting tomorrow morning, so I need my sleep."

She moves a hand toward my crotch, and I shift a leg to block it. "I'll make sure you have the best sleep of your life," she says, squeezing my thigh.

I jump out of the seat like a jack-in-the-box. "I've got to go to the toilet."

I am not repulsed by her attempt to seduce me, even though there is not a tingle of attraction. The more she drinks, the more melancholy she becomes, and I sense a deep inner sadness that undercuts her girlish display.

"What are you doing in Tonga, Pa'ula?" Sabrina asks when I return.

"I'm writing a book about man-eaters."

Sabrina throws a sly grin at me and I know her reply even before she utters it. "I'm the best man-eater in Tonga," she says. "Let me come to your hotel room and I'll prove it."

Fina snorts at this sally, and pulls from her tote bag a book. I see from the cover that it is William Mariner's account of his time with the Tongans. "Jesus, we were a bloodthirsty lot then," she says with a look that waivers between awe, a little fear and pride. "I read how Mariner described the cooking of human bodies. It was gruesome. But the Fijians were no different." She takes out a few letter-size sheets from the bag and a couple of books.

"The librarian helped me, but it's mostly about Fijian cannibals." she explains. "There's not much written about our cannibals."

Tonga and Fiji, close Pacific neighbors, have had a bellicose rivalry lasting for centuries, the battlefield now transformed into a rugby field, and there is often an intense dislike between them. So, I am not surprised when she seems to be smoothing over her own people's cannibalism, or perhaps diverting my attention toward the enemy.

"Listen to this," she says. "A Methodist minister, the Reverend Jaggar, entered in his diary in 1844, and I'm quoting him: "The men doomed to death were made to dig a hole in the earth for an oven. To roast their own bodies. Sern, the Bau chief, then had their arms and legs cut off, cooked and eaten, some of the flesh being presented to the prisoners. He then ordered a fish hook to be put into their tongues, which were then drawn out as far as possible before being cut off. These were roasted and eaten to the taunts of 'We are eating your tongues.'"

"Are you sure you'd trust a European missionary to tell the truth about cannibalism?" I ask.

Fina studies me with a look of amusement and a curl of her upper lip. "I've known plenty of *palangi* missionaries, and none have ever told me a lie. But you're the one who's supposed to be the expert on cannibals, and now you're questioning the clear evidence. How about this?" She reads from a sheet of paper containing an excerpt from a book.

> Eating the enemy's flesh was the Fijian warrior's most fervent insult. The purser on a US ship charting Fiji in 1840, William Speiden, observed that, "One man actually stood by my side and ate the very eyes out of a roasted skull he had, saying, '*Venaca, venaca,*' that is, 'very good.'"

"Yum, yum," Sabrina says in scorn, rolling her own pretty eyes.

"Yum, yum, your bum," Fina retorts. "I've eaten pig's eyes at feasts and they're pretty juicy. I'd bet human eyes taste much the same. Anyway, my father's father was a Fijian, from Lautoka, and so I've got cannibalism in my blood from both sides. It's the way you're brought up. I've read that Muslims won't eat pork, they believe it's unclean, but it's our favorite food. They'd probably throw up in disgust if they came to one of our feasts and saw us tearing into the roast pigs."

"Yes, but there's a big difference between eating another species and eating one of your own," I suggest.

"You think so? I bet that if you were born into a cannibal tribe, then eating humans would feel no different to you than eating a plate of prawns or a roast pig. That New Guinea tribe you told us about. Did they tell you what human flesh tasted like?"

"They said it was very tasty, especially the brains and the tongue."

Sabrina, pouting, thrusts out her tongue and waggles it. "You can eat my tongue, Pa'ula. Aneeeeee time you like."

Fina ignores the interruption. "See what I mean. We're all meat." She takes another photocopied page. "Listen, here's some more, about cannibalism in Easter Island. 'Every Easter Islander knows that his ancestors were *kai-tangata*, man-eaters. Cannibal feasts were held in secluded spots, and women and children rarely admitted. The natives said that the fingers and toes were the choicest morsels. The captives destined to be eaten were shut up in huts in front of the sanctuaries. There, they were kept until the moment when they were sacrificed to the gods. The Easter Islanders' cannibalism was not exclusively a religious rite, or the expression of an urge to revenge; it was also induced by a simple liking for human flesh that could impel a man to kill for no other reason than his desire for fresh meat.'"

Sabrina throws her hands back in mock horror. "Oh Fina, put a plug in it. We've come here to have fun and dance and drink the night away."

Fina abruptly stands up and takes my hand. "Let's go back to your hotel. It's quieter there," she says.

This beautiful Tongan girl has me confused. Does she sleep with men, or *fakaleitis*? Whatever, I did not want to put her to the test. "Thanks, but as I told Sabrina, I'd rather get a good night's sleep."

"Silly," she giggles. "I don't want to sleep with you. It's easier to talk there."

☠

Tavita has gone and we take another taxi, as decrepit as his, to the International Dateline. Although it is nearly two in the morning, the small bar by the swimming pool is still offering service, though the barman can barely keep his eyes open. He smiles when he sees Fina approaching, as if she is a regular customer. "It never closes," she tells me after ordering two bottles of local beer.

Fina puts one of the bottles in her mouth and, gripping the top between her strong white teeth, wrenches it off. The malty beer froths over the rim.

I remove my top with a bottle opener, not wanting her to break a tooth for my sake.

"I want to tell you about the *fakaleitis*. They might look happy, but they're some of the saddest people I've ever met."

"You seem pretty close to them, especially to Sabrina."

She shakes her head. "*Palangis* always get it wrong. It's not what you think. They're born like that, they really think they're girls, even when they are children. It's a Polynesian thing and many families raise them as girls. But, on the outer islands, they often get bashed by the men, and they run away to Nuku'alofa."

"How do you know the *fakaleitis*?"

"They find it hard to get a job, so some families take them in, give them food and a bed in exchange for doing the cooking, washing, cleaning. With all those muscles, they're hard workers."

"Does your family have a *fakaleiti*?"

"I live alone in our family house near here. My father is dead, and my mother lives in California with my elder brother and younger sisters. So, I always have a *fakaleiti* with me, to do the housework and for company."

Fina gains from being friendly with the *fakaleitis*, cheap labor for her house and protection from robbers or worse, and yet they seem genuinely fond of her and she of them. "Life is even sadder for the *fakaleitis* than for most of us, and yet they can still laugh, still joke, still make fun of each other," she had told me earlier in the night on the way to the bar. "That's why I like them around me."

She tells me about the bachelor crown prince, fifty-seven-year-old Tupouto'a Tupou who, she says, had a *fakaleiti* as his nanny when he was a child. "He refuses to marry and grant the kingdom an heir because, two decades ago, the king refused him permission to marry the commoner he loved. His friends call him 'Tippytoes.' You should ask him about Tonga's cannibals."

CHAPTER 15

The next morning, thumbing through the phone book, I find the listing for the Crown Prince's mansion. Although it is ten times bigger than a normal Tongan home there is just one number. The phone rings for a couple of minutes and then a man answers. "What you want?" he asks gruffly.

"Could I speak with His Highness?"

"He not awake."

"But it's already ten o'clock in the morning."

"He always stay awake all night, sleep all day. If you want say something to him, you go see his private secretary, Mele Vikatolia."

Vikatolia is the Tongan way of spelling Victoria, and Mele's parents must have had high hopes for a girl they named after the formidable queen.

"Where can I find her?"

"Reserve Bank building."

I know something of Mele Vikatolia Faletau. She is the star of one of Tonga's most notorious love affairs. Sent to Oxford by her wealthy parents, she spurned the expected choice of Law and graduated in Classics, learning to speak Greek and Latin. It was a brave choice and marked her as someone with a view of life broader than mere ambition. There must be no opening for a Latin speaker in Nuku'alofa.

Whatever dreams she might have had of a brilliant career, on returning to the Tongan capital she was chosen by the crown prince as his personal assistant. Her father was heard to moan, "I spent all that money on her education so she could end up making tea for Tupouto'a."

Mele was married with three children when she left her husband, the king's nephew, and moved in with one of Tonga's nobles, Ma'afu, a handsome, shaven-headed, hawk-eyed soldier, descendant of a famously bellicose warlord. It is a tangled affair, because Mele was married to Prince Mailefihi, son of the king's brother, Prince Tui'pelehake, while Ma'afu was married to Princess Taone, sister of Mailefihi. Mele and her prince are divorced, but Ma'afu is still married to the princess, even though he lives with Mele, and Fina tells me the *fakaleiti* gossip has it that he has had a child with her. The affair is rarely mentioned in polite society in Tonga, and never in the media, but everyone knows about it.

At the Reserve Bank I take the lift up to the crown prince's office, which spreads over most of one floor. Through a glass door I spy a huge room containing a massive desk and an expansive expensive leather lounge suite. Large vases filled with freshly cut tropical flowers decorate the desk and a coffee table. The tropical sun streams in through the ceiling-to-floor glass windows. It is the office of a person with a king-size ego.

The large room looks empty, but then I spy a tall, thin, imperious woman sitting at a small desk around the corner from the entrance. Despite her aristocratic appearance, she is hunched over a typewriter and typing a business letter. She eyes me suspiciously when I enter. "What can I do for you?"

Mele Vikatolia does not offer her name, but her Oxbridge accent, as precise as a cut diamond, gives her away. Middle-aged, she looks more like a Mediterranean than a Tongan with a slim, high-cheeked face that must have been very pretty when she was younger, but is now marred by a frown. Tupouto'a lives a helter-skelter life, often setting off at short notice on a jaunt overseas, and Mele must abandon her noble and dog the heels of the crown prince on his journeys. "She's sick of the life, she'd rather stay at home, but Tupouto'a won't let her because she's so efficient, and you can't turn down the crown prince," one of Fina's well-connected friends told me.

"I'd like to meet Crown Prince Tupouto'a please."

"What for?" Her voice is as sharp-edged as her cheekbones.

"I'm a writer and I'm researching a book."

She looks at me as if I am a loathsome microbe. "His Highness does not talk to the foreign media," she says and turns back to her typewriter. The hint is clear.

"I wonder if you'd put my request to Crown Prince Tupouto'a."

She looks up at me with angry dark eyes. "If you insist, but I know his answer. He never gives interviews."

After lunch I lie on the hotel bed, enjoying a couple of hours sleep. The siesta is one of mankind's most beneficial inventions, especially in a tropical climate, where the energy flows from you with each drop of perspiration. The phone rings just before 2 pm. "It's too hot to work. Want to go for a swim? I'll take you to my favorite beach."

Even though I am drowsy, only the Tin Man, lacking a heart, could resist such an offer. Fina arrives at the hotel soon after in an old Toyota, its black paint chipped, its hood dented. "It belongs to one of my friends," she says. "The beach is about half an hour out of town."

We drive along the narrow road that traces the island's south coast, skirting coconut, vanilla and pumpkin plantations, dotted along the way by roadside settlements containing a few wooden bungalows, but always at least one church. Only the street dogs are silly enough to be awake at this time. Even the pigs doze after their morning of food-grubbing, huddled in clumps of dark porcine flesh under trees or in the shadows of verandas.

Fina turns off the road and we bounce down a dirt track for about a mile. I can hear the roar of the sea, but cannot see it as she parks the car. She leads me through a break in a wall of black volcanic rock and there suddenly is her beach, a curve of white sand edged against a scoop of blue water and guarded, about fifty yards out to sea, by a ledge of pitted volcanic rock that rises a few feet above the ocean.

Wave after wave rushes in from the sea to slam against the ledge, exploding in a tower of spume. Blowholes in the rock's natural vents shoot spouts of water high into the air with each assault on the ledge. No one else is at the beach, and it cannot be seen from the hill because of the rock wall in front of the slope.

It is Fina's private paradise. My heart stutters as she begins to remove her dress and top, but calms when she reveals underneath a pair of baggy shorts and a T-shirt. "It's against the law to wear skimpy swimming costumes in Tonga, by order of the king," she says. "We are, after all, a Christian country," she adds, voice laced with scorn, "and so we mustn't stir up sinful passions by wearing bikinis at the beach."

My board shorts pass muster. The water is cooling and, as we sit in one of the many clear pools and let our arms trail under the tug of the gentle current, the waves, their sting plucked by the rocky ledge, foam about our necks and splash our faces. "I've been coming here since I was a small girl," she says. "My mother and two younger sisters emigrated to California to be with my elder brother who lives there. America refused to let me go with them because I had just turned twenty-one. But I don't mind because I've seen their beaches on TV, and there's nothing like this. I come here whenever there's a slack day at work."

Fina suddenly rises and strides back to shore. She spreads her towel on a stretch of dry sand shielded on three sides by a six-foot-high chunk of rock near the water's edge. "I wonder if you'd mind going to the end of the beach and staying there until I call you back."

I walk away, sand squelching between my toes. About fifty yards down the beach I turn to see that she has taken off all her clothes and now lies stretched on the sand soaking up the sun. I look away, guilty at spying her naked body. I lie down on the warm sand and rest my chin against a bunched towel, staring out to sea. The sun stabs my eyes, forcing me to close them, and I drift away.

My eyes spring open. How long have I been asleep? Where is Fina? Down the beach I see her standing knee-deep, scooping sea water to let it slip over her breasts so that it tumbles down into the curve between her thighs. In the full swell of her body, with her dark hair trailing down her tawny back, she looks like a Polynesian Venus emerging from the sea. Strangely, despite all the beautiful flesh on display, the moment is too pure for lust. And yet, I think, in another life, at another time . . .

I turn, stare out to sea and, when I hear footsteps approaching, pretend to be asleep. A toe nudges my ribs. "Come on sleepy, it's time to go back to civilization."

On the drive back to town I tell Fina about my encounter with Mele Vikatolia. "You want to meet the crown prince? Easy. Just go to the Nuku'alofa Club at six tonight. He's usually awake by then, and he often goes to the club to drink with his cronies."

"Can you come with me?"

She screws her face in disgust. "Women are banned from joining that club, or even entering it. There are about two hundred members, the high and mighty of Tonga, all men of course. As far as they're concerned women are only good for cooking and making babies."

☠

I shower at the hotel and put on a pair of neatly pressed khaki pants, blue shirt, dark red tie and polished black shoes in the certainty that Tonga's most exclusive club will have a strict dress code. I have no way of knowing whether the crown prince will turn up, but it is worth the try to see Tonga's next monarch up close.

Tupouto'a graduated from the military academy at Sandhurst in England and, deprived of a real army, has become an expert on Napoleon's war against Russia. At a large cost he purchased a cast of expertly crafted toy soldiers to re-enact that conflict, and it must be sad to watch him as he maneuvers hundreds of them around his cavernous living room, deprived of real soldiers to display his strategic skill. He attends parades in Tonga clad in army and navy uniforms so flamboyant that military attachés from friendly countries can barely conceal their smiles.

Although the Tongan navy consists of a handful of patrol boats, Tupouto'a flaunts a full-dress naval commander's uniform awash with gold braid, gold buttons, service medals and sword buckled to a gold-striped belt. He looks like an actor in a small-town performance of Gilbert and Sullivan, but one of his friends told me that, protected by the haughtiness of an upbringing in a royal family that rules with near absolute power, he does not care one wink what his subjects think of him.

The Nuku'alofa Club is across the road from the palace. It is a large pleasant bungalow of the type once favored by expatriate planters all over the British Empire. An L-shaped bar takes care of the drinkers, who can drift over to several full-size billiard tables. If the exertions there tire them, large padded cane chairs are dotted around the club. The overhead fans are so arthritic that their labored turnings produce just a whisper of a breeze.

A club servant signs me in, and I prop against the bar and order a glass of the local beer, Royal. It is just before six and there are only four other members in the club, drinking at the far end of the bar. I recognize Tupouto'a from his pictures. He is clad in white slacks and white shirt, and sits on a stool at the bar clutching a Royal. A trio of cronies stand around him chatting.

The crown prince resembles a Polynesian Colonel Blimp with a well-padded, big-hipped body and a balding pate looming over liverish features. His heavy-lidded dark eyes are magnified by the thick lenses in his glasses. He looks a decade older than his fifty-seven years and hobbles to the toilet, suffering famously from gout earned by decades devoted to hard living.

I slowly edge down the bar, taking about fifteen minutes to reach Tupouto'a and his group. "Mind if I join you?" I ask, giving no hint that I know that he is the crown prince.

"Of course," he says in the same fleshy voice as the king. It sounds as if his mouth is stuffed with cotton wool. At first I say little, leaving most of the conversation to the cronies. They address the crown prince as "Sir" or "Your Highness" and laugh at all Tupouto'a's jokes, agreeing with his every opinion. Most tip off the political spectrum at the far right. Typical is his sure-fire cure for Tonga's serious drug problem. (Marijuana is widely grown and cocaine easily available.)

"If a dealer is caught with drugs, you put a nine-millimeter pistol between his eyes and pull the trigger," he says with venom. His acolytes nod eagerly. "As for the addicts, what they need is a good thumping. Let the police thrash the habit out of them."

Despite this brutal prescription, I am intrigued by Tupouto'a's effeminate voice and limp-wristed manner. They also intrigue Tongans. "As a child, he had a nanny who was a *fakaleiti*, and she made a lasting impression on the crown prince," a friend of his told me. "The *fakaleiti* made all his clothes and they were very girlish, with plenty of frills and bows."

Tupouto'a has responded to the inevitable rumours with characteristic bluntness, denying he is a homosexual. As if to prove it, his sex life is legendary in Tonga. His favorite quarry in his more vigorous days were beauty queens and shop assistants. A girl sired by Tupouto'a with a commoner was raised in the palace by the queen and then married off to a younger son of a noble.

The crown prince is a chain smoker and, when he realizes that I do not smoke, points to a sign over the bar. "Non-smoking is permitted in this club," he reads out and laughs.

He is clearly a serious drinker because the beer flows freely as the hours drift by. Knowing a challenging night approaches, I make frequent trips to the toilet where, behind a closed door, I pen his quotes in a notebook. I am almost certainly going to get drunk in the line of duty, and will not remember much the next day.

Not once do his cronies challenge the crown prince's often ignorant views and to liven the night I decide to take him on. "Catholic countries are hotbeds of revolution, always rebelling against their betters, because it is an emotional religion and its followers let their hearts get the better of their heads in everything," he says deep into the night. "But in Protestant countries the people respect the rule of law and that's why they are saved from turmoil." The trio nod their heads as if he has uttered a proverb for the ages.

"When it was a British colony, the US wasn't a Catholic country and yet the people rose up and rebelled against their so-called betters in their desire for freedom," I counter.

The crown prince frowns at this *lèse majesté*. With a plump arm he waves away my point. "That's an exception," he grumbles.

The cronies grimace when I ask Tupouto'a about Tonga's cannibal past. "Read William Mariner's account of his time in Vava'u, it's all there," he says.

"Does that mean your recent ancestors were cannibals?"

"Yes, and so were yours, somewhere back in time."

His flunkys on cue chuckle at their master's wit. Tupouto'a exercises royal privilege by shifting the conversation to rugby. With my head growing thicker with each glass of Royal I look at my watch and see that it is already midnight. We have been drinking for six hours. Then, my memory goes blank. I wake up at seven in the morning in my hotel bed, sore-headed and perspiring, with faint glimmers of memory of the night before. The notebook, have I lost it? I spy it on the bedside table. Thumbing through it, I breathe easy, reading pages of quotes and descriptions of Tupouto'a at the bar. By the end, though, the handwriting looks as if it were penned during an earthquake.

Fina meets me at the International Dateline that night. "Your face is aglow," she says.

"No, it's the flush from a hangover that I can't shake off. I got drunk with the crown prince last night."

"Bravo," she cries, kissing me on the cheek. "The hunter gets his prey."

That night we go to the Blue Pacific, but I am exhausted in mind and body and we do not linger. She says goodbye to me in the lobby well before midnight. Tavita drives us to the airport the next morning. At the entrance to immigration Fina and I hug for several moments. "Have a safe trip home," she says. "Come back one day before I get too old."

☠

The old king has since died. Tippytoes has ascended the Tongan throne as Siaosi Tāufa`āhau Manumataongo Tuku`aho Tupou, otherwise known as King George Tupou V. His formal coronation was scheduled for August 2007 but was postponed after Tongans in their thousands rioted and torched Nuku'alofa's main street in November 2006, burning down most shops and houses. The cause: the royal family's reluctance to transfer real power from themselves and the nobles to the common people. Ironically, the old king sent Tongan troops to Iraq as part of US President George W. Bush's push by the "Coalition of the Willing" to democratize that country.

Suffer the Little Children

UGANDA'S LORD'S RESISTANCE ARMY

CHAPTER 16

The lights inside a crowded 767 come on without warning, nudging awake most of the passengers. We have been in the air all through the night and we blink sleepy-eyed as a fruity-accented African stewardess announces that the jet will soon land at Entebbe airport in Uganda. I have left my home in Sydney again on my quest, and what I am to look for here is cannibalism still practiced and on a massive scale, the work of a rebel group known as the Lord's Resistance Army, or LRA. I sense the encounter will be traumatic.

The ragtag LRA troops roam northern Uganda's countryside in small units, emerging unpredictably to torch villages, kill people and kidnap children before returning to the forest. Their terror tactics and the bloody clashes between the rebels and the army have forced 1.6 million people, or about 90 percent of northern Uganda's population, to flee their homes and become refugees in their own country. These "internally displaced" Ugandans have been ordered to settle for years in squalid government camps where malnutrition, disease, violence, and crime are common. As a result, the international medical aid group Médecins Sans Frontières has declared a state of emergency in northern Uganda.

Before leaving Australia, I spoke by phone with Jane Abachweezi, a Ugandan writer and translator, who has gone north several times to the killing fields. "In Acholi province, which borders Sudan, a rebel leader, Joseph Kony, and his troops have forced at gunpoint several hundred children as young as six to eat human flesh after they were made to butcher their victims, their playmates and friends," she told me. The horrific numbers have come from social workers and psychologists who've spoken with most of the fifteen thousand children rescued over the past few months by the army. I find it difficult to believe—ten, fifteen, thirty, that's possible, but several hundred children? "Yes, and it's in the upper limit of the hundreds," she replied. "When you go north you can talk to many of the children."

If so, Joseph Kony must be the most evil person now on the planet.

☠

Below the jet, the clouds drift apart to reveal an emerald-green stretch of hills, valleys and fertile farmland. The dense forest cut by red dirt tracks and with an occasional hilly bump reminds me of the outback of my homeland. This, though, is the heartland of Africa, for the past few years my second home.

Africa! Cannibals! This present journey was always inevitable: those two powerfully emotive words seem linked in the Western mind like pork and apple sauce. Cannibalism of one kind or another has been common around our globe through the millennia, and yet the classic Western image of cannibals is a terrified white Christian missionary in pith helmet crouching in a large outdoor cooking pot, the logs burning fiercely as wild-eyed African warriors in grass skirts dance about him shaking their spears. Their glinting eyes show their eagerness to tuck into their human meal.

In truth there is not one record of a missionary ending up in an African cook pot. The cannibals invariably ate one another. Henry Morton Stanley's account of the cannibalism he encountered while searching for Dr Livingstone in 1871 and 1872 was spine-chilling. In his classic book, *Through the Dark Continent,* published in London in 1878, the itinerant American journalist and explorer describes stumbling into a war party of warriors who had just captured eight enemy men in concealed nets, trapping them as meat, to be killed and eaten by the tribe. Stanley found that the captives were themselves cannibals, and they told him that "they ate old men and old women as well as every stranger captured in the woods."

Vladimir Putin, the Russian president, referred to Africa's infamous cannibal past at a conference in Britain in 2006 while defending his own soiled human rights record. Putin said: "We all know that African countries used to have a tradition of eating their own adversaries. We don't have such a tradition or process or culture, and I believe the comparison between Africa and Russia is not quite just."

Inevitably, one of the commissars for political correctness that we in the West suffer—in this case, Trevor Phillips, head of the United Kingdom's Commission for Racial Equality—huffily retorted, "What a preposterous thing to say. He is at best insensitive and at worst a downright racist." And yet Putin was much closer to the truth than Phillips, because throughout history and even into this century, Africa has been home to many cannibal tribes, probably more at any given time than on any other continent.

Herodotus, the Greek historian, quoted eyewitnesses of African cannibalism more than two thousand years ago. Much closer to our own time, explorers in Africa such as the gorilla-hunting French-American Paul Du Chaillu shocked their readers with descriptions of villagers commonly eating human flesh. In 1861 Du Chaillu wrote that he found human flesh to be the daily fare in a tribe named the Fan in Gabon, then known as French Equatorial Africa. The warlike Fan men certainly looked the part. They carried weapons at all times, filed their teeth to demonic-looking points and decorated their bodies with tattoos and war paint.

Du Chaillu observed that on entering a Fan village, "I perceived some bloody remains which looked to me human, but I passed on, still incredulous. Presently we passed a woman who solved all doubt. She bore with her a piece of the thigh of a human body, just as we should go to the market and carry thence a roast or steak." The Fan told Du Chaillu that they ate prisoners taken in war.

The king of the Apingi tribe gave Du Chaillu a grisly present, a slave bound by ropes. "Kill him for your evening meal; he is tender and fat, and you must be hungry."

In 1924, a British colonial official in Uganda, John Roscoe, noted of the locals: "There is not the slightest doubt in my mind that they prefer human flesh to any other." He told of a young chief visiting his camp who cut the throat of a little slave girl he had brought along as food, and began to cook her until he was stopped by one of Roscoe's soldiers. "He had a bag slung round his neck which, on examining it, we found contained an arm and leg of a young child."

William Arens cast doubts on such reports. He made a valid point when he wrote in his book, *The Man-eating Myth*: "As usual, he [Stanley] did not observe cannibals in the act, but relied primarily for his information on the word of Arab slavers in the area, who had a vested economic interest in discouraging European encroachment in their preserves, since it posed a threat to the lucrative trade in human beings."

He added another telling point: "David Livingstone considered the evidence for cannibalism among an African group he was familiar with, and concluded with the terse comment: 'A Scotch jury would say, Not Proven.' Since Livingstone spent a good portion of his adult life in east and central Africa ministering to what he considered to be the spiritual and physical well-being of natives he lived and died among, his experienced testimony should be of some value."

But Livingstone's experience was confined to only a relatively small proportion of the thousands of tribes spread across Africa at the time he was there. Many of the cannibalism reports came from West Africa, home to people with vastly different cultures from those among whom Livingstone lived and journeyed and preached.

What Arens surely cannot contest is compelling modern evidence of cannibalism in Africa. Some recent African leaders have even hungered after human flesh, among the most recent being the Sierra Leone militia leader Samuel Hinga Norman, whose fighters practiced human sacrifice and cannibalism. Hinga Norman led the feared Kamajor traditional hunters, the backbone of the government's Civil Defense Force, during the country's bloody civil war from 1991 to 2000. At his trial for war crimes before a United Nations–backed tribunal in Freetown early in 2007, the prosecution quoted witnesses who described Hinga Norman and his fighters, high on drugs, parading severed heads and eating the roasted flesh and intestines of their victims. Hinga Norman died in captivity in February 2007 while still on trial.

Then there was Jean-Bedel Bokassa, who had a taste for young girls, literally, when he ruled the basket-case Central African Republic from 1966 to 1979. Bokassa gave the world a few easy laughs in 1977 when he emulated his hero Napoleon by crowning himself emperor at a stadium in the ramshackle capital, Bangui. The lavish ceremony cost more than fifteen million dollars, which must have rivaled the country's GDP at the time. Foreign correspondents in Bangui for the coronation picked up rumors that Bokassa was a cannibal whose favorite meal, when he couldn't get young girls, was fricassee of political opponent. He was also rumored to feed enemies to the crocodiles in the palace pond.

Two years ago, when I was in Bangui for a few days, I paid homage to Bokassa's enormous coronation throne, fashioned in the image of a rampant eagle and once swathed with twenty-four-karat gold. The gold had long been hacked away, and the rusting throne was shoved into the corner of an open basement in the now abandoned stadium, as decrepit and discredited as Bokassa's memory.

I tracked down one of Emperor Bokassa's cooks, a grizzled old man whose mind was still sharp, and asked about the macabre rumors. "Yes, Bokassa liked eating human flesh," he admitted. "It was no secret in the palace, we all knew, and he'd boast to us that there was no better way of triumphing over an opponent than by eating him. Though he never had me cook any human flesh, I sometimes saw the body parts. He had the remains cut up at his farm near Bangui."

Bokassa was a cannibal with a demonic sense of humor. William Dale, the US Ambassador in Bangui from 1973 to 1975, wrote that during his time there, "young girls began to disappear, one by one." One Sunday morning in 1975, the Swiss pastor at Bangui's Lutheran church asked to speak in private with Dale. While visiting Bokassa's farm, he told the ambassador, he saw two schoolgirls roped to a tree.

Aware of the persistent rumours of Bokassa's taste for human flesh, Dale remembered that when he first met the Emperor to present his credentials, the ruler remarked that "his grandfather had been a cannibal and sometimes he believed he had tendencies in that direction." Dale says Bokassa also told ambassadors at a palace dinner that "he sometimes wondered whether his ambassadorial guests realized exactly what they were eating."

Dale made a gruesome conclusion. "The ghastly thought occurred to me that perhaps Bokassa was actually dining on parts of the school girls, and was possibly serving them as a routine matter to his dinner guests." French investigators called in by the military leader who overthrew Bokassa found identifiable human parts, revealed as chunks of butchered missing schoolgirls, hanging in Bokassa's huge meat locker at his farm.

Not long after, a thousand miles to the east of Bangui, Idi Amin, Uganda's insane and mercurial dictator, claimed that he too had eaten human flesh but found it too salty, and this revelation was reported with relish by media around the world. Bokassa was an avid cannibal, and the best you could say for him was that he was faithfully carrying on the family tradition. But with Amin, you never knew whether he was serious or joking, unless confronted with the grisly evidence, and more than a hundred thousand bodies, many clogging Uganda's rivers and lakes during his bloody rule, proved that at the least he was a mass murderer on an industrial scale.

One witness, his guard, Abraham Sule, did see Amin "put his bayonet in a pot containing human blood and licked the stuff as it ran down the bayonet . . . Amin told us, 'When you lick the blood of your victim, you will not see nightmares.'"

Amin was recently depicted on screen in *The Last King of Scotland*. The American actor Forest Whitaker won an Academy Award as best actor in 2007 for his simplistic portrayal of Amin—one that swung between easily contrived bombast and belly-laughs, with little shading to give depth to the deeply evil character of the dictator. He explained to a reporter from the London *Daily Telegraph* his sympathy for Amin: "As an African-American, I relate to people trying to be who they are. Amin was trying to give pride to his people." Elsewhere, he was quoted by a South African news outlet as saying that Amin was not a nice man, but "he was not a monster." It's enough to make me wish I could tell Whitaker to his face to shut up and just read his lines.

Amin, the son of a sorceress, was certainly aggressively evil, the most brutal tyrant to that point in Africa's history. But a more gruesome monster lives on the loose in Uganda's far north.

☠

Entebbe, though decrepit and staffed by indolent officials, is considerably better than the usual ramshackle African airport, dark, dingy and crawling with touts offering to get you swiftly through the tedious drawn-out formalities if you pay them a fee in US dollars. That's not to say the money won't be well spent. At Niamey airport in Niger, with much experience of the scam, I paid twenty dollars to a middleman who ushered me toward the customs bench with a wink at an official. I was waved through without a search. A gaggle of French tourists whom I had seen refuse to pay the bribe were having their luggage contents strewn across the benches by vengeful customs officers.

Some African airports are just plain rough. The first time I entered the Congo through Kinshasa airport I had to fight off an African gentleman, clad in a sharp business suit and flaunting a gold watch, when he jumped several spots and then attempted to push me out of place near the front of the immigration queue. I stood my ground and watched other passengers manhandled by queue-jumpers fight back with punches and kung-fu kicks until the arrival hall looked like a fight scene from a Jackie Chan movie. The airport security guards stood aside and let the passengers sort themselves out. The next time I arrived at Kinshasa airport I gladly paid the fifty dollars demanded by my contact there to pass swiftly through immigration and customs, leaving in my wake another flurry of punching, thrusting, kicking passengers.

Entebbe, with its British colonial past, thankfully has orderly queues, although the airport was the scene of a heroic incident in 1976, a rescue mission by Israeli commandoes. They thumped down on the strip at midnight in a C-130 military plane and, with machine guns blazing, rescued 100 Israeli passengers who were being held hostage at the airport. Their hijackers, Palestinian guerrillas, had the approval of Amin.

☠

Waiting outside and holding aloft my name printed in marker pen on a pad is a round-faced, bright-eyed girl who looks to be in her early twenties. She is clad in a Congo-style ankle-nudging cotton dress that fits tightly about her neatly rounded thighs, and a short-sleeved top printed with a spray of red orchids that clings to her high firm breasts. She has woven her hair into strands festooned with colored beads. Unlike most of the women at the airport who are laden with fat and boasting the enormous bottoms that most African men are said to lust for, she is sleek and silky.

"Welcome to Uganda, Paul" she smiles, the Technicolor beads rattling as she tilts her head a little. "I'm Jane Abachweezi."

She has a taxi waiting outside and we head for Kampala, Uganda's capital, twenty miles away. The bitumen road, a procession of ragged potholes, is hemmed in by tumbledown shacks and open-air markets thronged with people. Bloodied goat carcasses hang on hooks next to stalls selling second-hand clothes from the West. Shadowing us on the right is Lake Victoria, the headwaters of the White Nile, a vast inland sea with a surface area of 26,600 square miles. Wind stirs the murky water into choppy waves, and a squadron of a dozen pelicans hurtles in single file low across the lake.

Jane and I have already discussed African cannibalism on the phone. As we drive, I show her an Associated Press story describing how rebel soldiers in the Congo's nearby Ituri province killed and ate pygmies, grilling their bodies on a spit and boiling two girls alive for food as their horrified mother watched. The story quoted the commander of UN forces in Congo, General Patrick Cammaert, as stating in March 2005 that the cannibalism was uncovered during testimony gathered from witnesses kidnapped by the rebels along the country's turbulent eastern border, abutting Uganda.

One witness, a pygmy woman named Zainabo Alfani, said rebels forced her to watch as they killed and ate her two children in June 2003. "In one corner, there was already cooked flesh from bodies and two bodies being grilled on a barbecue and, at the same time they prepared her two little girls, putting them alive in two big pots filled with boiling water and oil."

Jane looks straight ahead, her lips drawn tight. She sighs and murmurs, "My poor Africa." There is more horror. The BBC quoted the UN summary thus: "In one case, investigators heard how a young girl was cut into small pieces by the soldiers and then eaten. Other examples include hearts and other organs being cut out of victims and forced on their families to eat."

On two recent trips to the Central African Republic I spent several weeks with the BaAka pygmies in the Dzanga-Sangha national park. My good friend Wasse, the clan's best hunter there, told me that the taller Africans, known as the Bantu, had for centuries called the pygmies Babinga, a name that marks them indelibly as subhuman. He said this allowed the Bantu to kill and eat pygmies without guilt in the belief that it was not much different from eating forest animals. In revenge, Wasse told me with a wry smile that the pygmies called the Bantu *ebobo*, or gorillas.

Jane nods when I tell her about this. "We have pygmies here in Uganda and they seem just like little people to me, but many believe they are more like animals than humans. For centuries our people never saw the pygmies because they lived deep in the rainforest, and when they traded with us, they'd secretly bring bush honey and antelope meat to the jungle's edge and leave it there. Our people would take it, and leave for the pygmies cassava, which they didn't know how to grow but liked very much. It took hundreds of years for our people to lure many of them out of the jungle to work as slaves in our fields in return for a daily ration of cassava. That's why most of our people still think of pygmies as less than human, and why some can even eat them."

She stares out the window for a few minutes as we zoom past a landscape lush with trees and bush, with the shantytowns encrusted on the foliage like scabs on sleek skin. "Among my own people, the Banyoro, we have cannibals," she says as the traffic thickens to a crawl with our entry into Kampala's suburbs. "My hometown, Hoima, is near the Congo border, near where the rebels ate the pygmies, and some Banyoro clans there eat human flesh even today."

"Have you seen them doing it, or are these just tales you've heard?"

She snorts in disgust at my skeptic's retort. "Why would I tell you such a terrible thing if it weren't true? About two hours' drive from Hoima, there are villages we don't dare go near because the people there are cannibals, everyone knows that. Once, some of my university friends bravely visited there to see if the stories were true. They found parts of human bodies such as arms and legs hanging on lines to dry as if it were meat from animals. One of the boys entered a hut and sat down on a stool. The stool had been placed over a deep hole hidden by a mat. The moment he sat on the stool, he disappeared into the hole. My other friends ran away, terrified, and the boy has never been seen again. They think the villagers killed and ate him."

It sounds like a grisly Ugandan version of an urban myth, but Jane is convinced that the story is true and that her friends were not fooling her. She says she knew the boy, and hasn't seen him since the visit to the village. Even though I have known her for less than an hour, she seems sensible and intelligent, someone who would not feed me lies or half-truths about cannibalism just to please me. Yet I still doubt the story.

"If everyone knows the villagers are cannibals, and that people like your friend disappear there, then why don't the police go and arrest them?"

She rolls her eyes in scorn. "You Westerners, you think it's that easy. The villagers are masters of *ju-ju*, black magic, they have powerful witchdoctors, and even the police are afraid to make them angry for fear of what they'll do to their families."

☠

Kampala perches on seven hills. The road into the town center dips and soars, hemmed in by tin-roofed shacks with dirty curtains for doors and snot-nosed children sitting or playing outside. Everywhere you look you see poverty.

Uganda has been led for more than two decades by Yoweri Museveni, a former Marxist guerrilla leader turned ardent capitalist who liberated the people from a ruthless dictator, Milton Obote, in 1986. The ramshackle economy is testimony to his gross incompetence in running a country. Like Mao Zedong, Fidel Castro, Robert Mugabe and other illustrious twentieth-century guerrilla heroes, once Museveni tasted power he refused to give it up. Having won the right to rule by the power of a gun barrel, all he seems to know is how to use brute force to crush opposition when he is criticized.

"Most of us who live in the cities hate Museveni," Jane tells me, "but he's popular in the countryside. Even so, he needed to rig the last election in 2006 to hold power."

Museveni's opponents have been jailed or have disappeared into the maw of his secret police. Its members are expert assassins and torturers, and many of their victims have never been seen again. The US State Department has condemned Uganda's human rights record as poor. "Security forces committed unlawful killings," a recent report noted, listing numerous murders of civilians by soldiers. "Torture by security forces and beating of suspects to force confessions were serious problems." The report added that prison inmates have complained that security forces tortured them using snakes and crocodiles. At least Museveni didn't have them fed to the crocodiles like Emperor Bokassa's victims.

Despite the widespread barbarity, Western governments often laud Uganda as one of Africa's success stories, hailing Museveni's willingness to bow to World Bank and IMF dictates on free trade and privatization, as if they alone could save the country. In a quick sweep around Africa in July 2003, US President George W. Bush visited Kampala to laud the soldier-politician Museveni as "a strong advocate for free trade," but carefully neglected to add that he is both a killer and a tyrant.

Even the so-called booming economy touches few Ugandans. Kampala grabs most of the country's wealth, and boasts glinting office towers, fancy restaurants and flashy cars, a dolce-vita lifestyle that—as poor as much of the city still is—stands in stark contrast to the impoverished countryside.

"You think you've seen poverty in Kampala, but you've not seen anything yet," Jane says when I remark about the contrast. "The moment you leave the capital, as you'll see, the true poverty of Uganda smacks you in the face."

Uganda, with a per-capita income of just US$270, is one of the world's poorest countries, with 38 percent of Ugandans living below the national poverty line of US$1 per day. To survive, Uganda has become a beggar nation with European donor countries gifting Uganda 50 percent of its annual two billion dollar national budget. This is misguided largesse because it allows Museveni and his clique to steal from foreign aid budgets and from the roughly one billion dollars yearly that the country produces for export.

"They splurge the loot on fast cars, mansions, shopping trips to the US and Europe, Swiss bank accounts and large-bottomed mistresses," Jane growls. I can't keep myself from asking why large bottoms are so seductive. "Because our men believe them to be more juicy than girls like me," she says with a mocking laugh as she pats her round taut behind, either belittling her own lack of bulk or the men for their perverted taste.

☠

We pass at my request through the suburb of Wandegeya, home of the University of Makerere, once one of Africa's best universities but now a disintegrating hulk of its former illustrious self, just like Uganda itself. The writer Paul Theroux was a lecturer here back in the 1960s, along with V. S. Naipaul, and when he returned a few years ago was shocked at the deterioration.

Theroux wrote in *Dark Star Safari: Overland from Cairo to Cape Town*, published in 2003, that "the library—always a good gauge of the health of a university—was in very poor shape, unmaintained, with few users in sight and many empty shelves. What few books remained on the shelves were dusty and torn. I guessed that the books had been stolen. There were no new books. What had been the best library in East Africa was now just a shell."

Jane has a love and an aptitude for mathematics, and studied computer technology at Makerere, coping with the lack of hardware by sharing the classroom computers with other students. After graduation, though, she found it considerably more lucrative and much more fun to be a writer and work as a translator for visiting foreign reporters. "A doctor gets the equivalent of about three hundred dollars a month, and I'd be lucky if I earned two thirds of that as a computer expert," she explains. "In Kampala, that's a subsistence salary, even for a single person."

I have the window down, and a scream suddenly shatters the everyday buzz of the street. From an alleyway, a barefoot young man dashes across the road, his bulging eyes pulsing with fear. Our car jerks to a halt as he skips around the hood. His grubby polo shirt has been half torn from his back, and blood spurts from a deep cut on his neck. He sprints up a side street, and disappears. Moments later a shouting mob of about fifty men and women brandishing machetes, clubs and rocks race past us in pursuit. People call out the way he went, and some pick up rocks and dash forward to join the mob.

"Ah, he's a thief and he's soon going to die," Jane says with a grim smile as the car stutters to a start and heads up the hill. I'm shocked into silence by the brutality that thickens her voice. She gives me a cool hard look. "You're upset, aren't you? You think I'm just a black savage, but unless you live among us you'll never understand how we can do this, and even enjoy it. Once those people catch that thief, they'll stone him half to death, and then they'll set him afire and burn him alive."

"You think mob killing is justice?"

"Of course. I've been at three killings where I live, and each time when I heard the commotion I grabbed firewood and ran after the thief. Each time I was very happy when we burned him alive."

I am stunned not only by the murderous mob I have just witnessed, but by Jane's glee.

"These thieves rob us of everything we have," she explains. "Most of us don't get paid much, and so we have to save for years to buy TVs, refrigerators and other luxuries." Her eyes glower with anger. "These thieves take them from us in just a few moments, usually at gunpoint. They even rape girls if they get the chance. But if we catch a thief and take him to the police, we know that within hours they'll let him go after he's paid a bribe, and he'll be after us again soon enough. Our police are very corrupt. So, we hand out our own instant justice to thieves, and whether you like it or not, that helps keep our homes and girls safe from them."

"Don't the police intervene to stop the murder?"

"If they know about it they will, but by the time they reach the spot the thief is usually dead."

"I hope his killers don't eat him."

Jane stares at me for a few moments, puzzled, and then laughs. "I know what you mean. One way to keep a sane mind when thinking of all this horror is to turn it into a joke. That's Psychology 101."

☠

The city center of Kampala sits atop one of the hills. The sky above is Hitchcockian. Hundred of gargantuan marabou cranes speckle the sky as they wheel, twist and turn high above, riding the air currents with barely a flap of their enormous extended wings as they search the ground below for dead animals, garbage and anything else they can eat. Jane tells me that many Kampalans claim the storks fly through open windows and snatch babies from their cradles, to eat them at leisure in their nests. Unlike her hometown cannibal stories, this is one urban myth she refuses to believe.

The giant storks have a primeval look, like creatures from a far distant time, with their huge long beaks, squinting black evil eyes and unimaginable bulk. They seem far too big to fly. Close to ground level, hundreds more silently stand as still as statues on the tops of the many leafy trees lining the town's streets, their white feathers splattered and soiled from grubbing in Kampala's garbage dumps.

Later in the day, at the Africana Hotel, I relax on the room balcony sipping an after-lunch beer and admiring a view across a park to the city, fresh and gleaming from this distance. The balcony suddenly darkens as an enormous marabou hurtles toward me like a dive-bomber. A moment before the stork would have crashed into the railing, it dips sideways and shears past, its wing tip almost touching me.

☠

I have just one day in the Ugandan capital. After my brush with the bird, I spend most of the time reading a riveting book by Henry Kyemba, one of Idi Amin's best friends and a long-time minister in his government. Kyemba fled to England in 1977 when he sensed that the dictator was planning to kill him. In exile, he wrote *A State of Blood*, published in the same year. The dictator had a fondness for comically grandiose titles: His Excellency President for Life Field Marshall Al Hadj Doctor Idi Amin Dada, VC, DSO, MC, Lord of all the Beasts of Earth and Fishes of the Sea and Conqueror of the British Empire in Africa in General and Uganda in Particular. But you laughed at him at your peril. Almost every page of Kyemba's book recounts some horrific single or mass murder committed by Amin. He was the Pol Pot of Africa.

One of the most tragic episodes was the revenge the Lord of all the Beasts of Earth took on his second wife, Kay, a clergyman's daughter, who was a student at Makerere University when he married her. Divorced by the dictator, she took a local doctor, Mbalu-Mukasa, as her lover. Kyemba recounts that Kay became pregnant by the doctor who then performed a secret abortion on her, but she died under the anesthetic. The doctor cut off her arms and legs and bundled them with the torso in a burlap bag in an attempt to dump the body in a river, but police, alerted to the death, found the body in the boot of the doctor's car.

When this was discovered, a furious Amin ordered doctors at the Kampala hospital to stitch the arms and legs back onto her torso and place her on a table in the mortuary used for post-mortems. He then led the three children he had with Kay, aged between four and eight, into the mortuary to view the butchered body. According to Kyemba, who was there, Amin shouted at his children, "Your mother was a bad woman. See what has happened to her." The doctor escaped Amin's murderous wrath by committing suicide.

The Last King of Scotland, however, had Amin's fictional personal doctor, a young white man, at the mortuary when Amin stormed in with the children. The film followed the fictional book of the same title, on which it was based, in replacing the Ugandan doctor with the protagonist, the white doctor, as Kay's secret lover and witness of Amin's rage at the mortuary.

Though widely regarded as a buffoon, no one who knew him thought Amin was stupid. One of his final acts as Uganda's dictator was to caution Prince Charles not to marry Diana. "You will live to regret this," he warned.

Whether Amin was honest when he claimed he had eaten human flesh is impossible to know. Even if he were lying, he surely recognized the profound psychological impact the claim would have on his people. Cannibalism has long been wielded as a psychological weapon in Africa.

Sir Richard Burton, the great nineteenth-century explorer, wrote in his epic *The Lake Regions of Central Africa,* published in 1860, of a fierce tribe called the Wadoe. They drank out of unpolished human skulls, like the Aghori sadhus, and interred a male and female slave while still alive in the graves of great chiefs to make sure they continued to serve their master in the afterlife.

Burton witnessed them win a battle over their enemies, a tribe called the Wakamba. He wrote of the Wadoe: "Fearing defeat [they] proceeded in the presence of the foe to roast and devour slices from the bodies of the fallen." The result was a Wadoe victory: "The Wakamba could dare to die, but they could not face the idea of becoming food."

Joseph Kony, the rebel leader in Acholi province, also knows the psychological power of cannibalism. As I leave the hotel soon after dawn the next day to make the four-hour drive to the Acholi capital, Gulu, I brace myself for what threatens to be a harrowing experience.

CHAPTER 17

The car speeds north along the disintegrating bitumen, heading for the garrison town of Gulu, headquarters of the Ugandan government's fight against Joseph Kony and his macabre Lord's Resistance Army. For two decades Kony and his killers have terrorised northern Uganda, abducting from villages across the province more than twenty-five thousand children according to UNICEF, and turning many of the girls into sex slaves for the officers and the boys into merciless child soldiers. "The LRA has so terrorized the Acholi people that more than 90 percent of the population of almost two million have abandoned their villages and fled for protection to government concentration camps, where they've lived for many years in misery," Jane had told me as we headed out of Kampala three hours earlier.

Now, she is asleep in the back seat, her head slumped on her chest. Today, she has on blue jeans, a loose T-shirt, and sneakers, her work clothes. Rarely able to sleep in a car, I stare out the window at the lush green countryside and wonder how the people can be so poor in such a fertile land. She was right: The poverty I saw in Kampala was nothing compared to what I have seen in the tumbledown towns we have passed.

I nudge Jane awake as we descend a steep hill leading to a bridge fording the Karuma Falls. A wild torrent of roiling foamy white-capped water charges toward the bridge, crashing over a waterfall and then smashing past huge boulders strewn in its path. "It's the White Nile," she says sleepily, "and it's still got a few thousand miles to go before it reaches the Mediterranean. But we're about an hour away from Gulu."

She looks out the window as we climb a heavily wooded hill looming over the falls. "A few years ago LRA rebels ambushed a bus here and killed everybody on board," she says, explaining why the driver refused to stop on the bridge for me to take a picture of the waterfall. "I still remember the TV pictures, the bodies lying in pools of their own blood. We're now in dangerous territory, because the LRA rules here, especially at night."

The car trundles on, bouncing over the many potholes in the road, and I bury my nose in a book about the genesis of the Lord's Resistance Army. In 1986, Museveni's guerrillas overthrew the repressive government of Milton Obote and took power in Kampala after a bloody five-year struggle. Battle-weary, they were not prepared for the sudden rise in the north of an avenging thirty-year-old Christian prophetess. The bellicose northerner Acholis have never got on with the far more numerous southerners, whose governments often persecuted them with mass killings, assassinations of their leaders and throttling economic bias. Idi Amin had purged his army of Acholi troops, murdering many of them. The prophetess, Alice Lakwena, claimed she called her compatriots to arms after discovering in her village near Gulu that Museveni's troops had "rounded up hundreds of young boys and girls and slaughtered them."

Lakwena declared a holy war on the south. Her Holy Spirit Movement spearheaded the assault. According to Heike Behrend's fine biography of the prophetess, *Alice Lakwena and the Holy Spirits*, published in 1999, Alice's family name was Auma, but she claimed to be possessed by a powerful sacred spirit named Lakwena, which means "spiritual messenger" in Acholi. None of the fifteen witchdoctors her father took her to could cast out the powerful Lakwena, and Alice took on the spirit's name as her own.

Incandescent with charisma, she lured the former chief of staff of the Ugandan Army and the former minister of education into her camp even though she could arm her ragtag militia only with sticks and stones and a few assault rifles. She sprinkled holy water on her soldiers to invest them with supreme power when going into battle, and gave them magical stones she claimed would turn into grenades when thrown at the enemy. Lakwena also ordered her troops to remain chaste to ward off the bullets with an invisible shield of purity. She would then claim that any of her soldiers who died in battle clearly had sinned beforehand. Those who survived praised her power to protect them.

The highly motivated rebels stormed south, defeating several brigades of Museveni's tired troops in major battles. Within a year they reached the strategic town of Jinja on the Nile's banks, just fifty miles from Kampala, and seemed set to seize the capital. But in October 1987 the army brought in Katyusha rockets, up to forty-eight in each bundle mounted on a heavy truck, and these blasted through the rebel ranks, whose sole strategy was to swarm over the enemy troops en masse. The rockets turned the tide, and with her rebel army literally torn apart, Lakwena fled to Kenya where she died in exile in a refugee camp in 2007.

But Lakwena's rebellion spawned the Lord's Resistance Army, with Joseph Kony, her reputed cousin and a former altar boy, grabbing the leadership of the Holy Spirit Movement two decades ago. He invested it with a powerful melange of black magic, hocus pocus and Catholic hymns. Kony's surname in the Acholi language means "to help," and flourishing a Bible gave him credibility among the Acholi, a highly religious people. He demanded that his followers keep strictly the Ten Commandments, and added an eleventh, declaring it a sin to ride a pushbike, the most common means of transport in rural Uganda, on Friday. The LRA punishes this sin with execution. Adultery and premarital sex are also mortal sins among the LRA, and Kony forces the guilty pair to face a firing squad.

Kony and his Acholi rebels are still in the bush. According to the UN more than one hundred thousand people have been killed in the conflict, which has cost the impoverished nation at least $1.5 billion. The US State Department classifies the LRA as a terrorist organization, and the United States has provided more than one hundred million dollars to Uganda for food and other forms of assistance including support for reintegrating former child soldiers and formerly abducted persons into Ugandan life.

At a village near Karuma I buy from a street hawker a copy of the national newspaper, the *Daily Monitor,* and find that the day before LRA rebels had crept into a village just two miles from the center of Gulu and hacked off the lips and ears of a man they believed to be a soldier in the national army. They caught him wearing gum boots, standard issue for the army, but also worn by farmers when they work in muddy fields. "Mutilating the face like that is a trademark LRA tactic," Jane tells me. "It gives warning to anyone who sees the person not to oppose the LRA or suffer the same fate."

☠

As we near Gulu, the villages begin to disappear, replaced by vast assemblies of mud-hut villages, concentration camps guarded day and night by soldiers wearing gum boots. A town tormented daily by fear of the marauding LRA, Gulu is home to the battle-hardened 4th Division and soldiers, assault rifles slung on their shoulders, stroll along the potholed footpaths or drive by in pickup trucks. The road we follow into town is lined by crumbling shop houses and leads to a shantytown marketplace.

We book in at the Acholi Inn, headquarters of the so far futile attempts by the Ugandan government to bring the LRA to heel with a peace agreement. It has the familiar indolent air of an up-country one-star hotel in eastern Africa, and when we ask to have lunch in the otherwise deserted restaurant, a slow-moving middle-aged waiter dutifully places menus in front of Jane and me. Hungry from the long drive, we each choose a full meal—soup, main course and dessert—but when the waiter returns to take the orders he laughs mirthlessly.

"Only baked beans and toast for lunch today," he chortles.

"Then why give us the menu?" I ask.

"Only baked beans and toast for lunch," he replies, and shuffles off to the kitchen.

The Acholi Inn is owned by a colonel who headed army intelligence in Gulu. He has made a pile of money out of the LRA rebellion, but is not unusual in combining war and business. Ugandan senior army officers indulge in many kinds of money-making schemes, even when there is a clear conflict of interest as here. Jane tells me that army officers on training schemes use the colonel's hotel, as do government peace negotiators who stay for months at a time. Still, his ownership of the lucrative hotel is considerably more sanitized than senior officers who profit directly from the long-running war. "There've been stories in the newspapers of Museveni and his senior officers not wanting to end the conflict because they cream off a lot of money from the military aid given them by the US to fight the rebels," Jane tells me.

After a short siesta, we drive down a dusty road shaded by leafy trees to the Children of War Rehabilitation Centre, where most of the LRA women and children captured by the army are taken to be questioned and counseled at the beginning of their difficult return to society. The car stops out-

side a high shuttered gate and walls studded with broken glass. I suspect
the glass is to stop people getting out, not getting in. You need special
permission to enter here, and we wait in the mind-numbing heat while the
guard checks our credentials.

As the gates swing open the hair rises on the back of my neck. Hundreds of children and women are crammed into a small compound of about
half an acre scattered with bungalows and UNICEF tents. Some children
play catch; others perform traditional hip-twitching dances to the throb
of drums. Hard-eyed boys in their early teens scurry around the camp on
crutches, most missing part or all of a leg, probably from a land mine or a
badly treated bullet wound in an LRA camp.

I wince as a young girl hobbles past. Her coordination seems awry, perhaps because the right side of her skull has been hacked away by an ax or
machete. It has healed in a crude way with the scar tissue stretched across

*At the Children of War Rehabilitation Centre at Gulu, where children rescued from
the LRA by the army are taken after capture for rehabilitation. Thirteen-year-old
Steler Layet stares blankly into space as her mother Jerodina hugs her. Steler was
rescued from the LRA just three days earlier after being abducted and held for three
years by the LRA. Her family rushed to the center to be with her, but feelings of
shame over what she must have done while with the LRA mar the meeting.*

a spongy indentation the size of my palm. Dozens of young women sit in the shade cradling babies.

Apart from the battle wounds and amputations, it looks like any one of the scores of refugee camps I have visited around the world, a common sour smell of the inmates' misery and hopelessness mingling with the over-run of a rough-and-ready sewage system unable to cope with the numbers. But there is a special sense of terror that thickens the air here because these children and women have journeyed to the outer chambers of hell and returned. They have experienced horrors so nightmarish that when I see them up close, their eyes seem thrust deep into their sockets, seared with anguish, despair, pain and shame.

"You can tell the most recent arrivals," Jane says, pointing to a huddle of about forty women and children enveloped in a shadowy silence. They have haunted stares and bone-thin bodies disfigured by skin diseases and sup-purating sores. She has visited this center before, and is familiar with their fate. "They'd have been captured by the army just a few days ago. They were probably attacked in their rebel bush camps by helicopter gunships shooting to kill anyone below—man, woman or child."

In a small cramped office near the gates, Michael Oruni, the Children of War Rehabilitation Centre's administrator, and himself a psychosocial worker, rises to meet me. He is an Acholi, tall with jet-black skin. "We've had fifteen thousand child-mothers and children pass through here over the past decade, all victims of Joseph Kony and the LRA. They've all had horrible experiences, far beyond the worst nightmare you can imagine."

We are joined by Jacqueline Okongo, a graduate social worker and one of the trained counselors who daily attempt to ease from the kids their horrific tales as part of their therapy. Jacqueline is a plump young woman with a kindly smile, and I can imagine a child who has just escaped from the LRA falling into her comforting motherly hug. "All the children have been deeply traumatized by their time with the LRA," she says as we walk through the camp. "But the most deeply scarred mentally are those who had to murder and eat other children. To frighten the children from trying to escape, Kony ordered that any who tried must be killed immediately by the other children, and then cut up and eaten by them."

I ask her how common was this demonic punishment. "According to the records we keep, Kony and his troops forced at the very least many hundreds of his little captives to eat human flesh."

At the Children of War Rehabilitation Centre counselor Jacqueline Okongo sits with a child-mother, Beatrice Ochora. who had been abducted as a child from her village by the LRA and had just been rescued by the Ugandan army. She had a child with one of the LRA officers, having been forced into a marriage with him.

It is the children who freely admit this to the three counselors at the camp who have the best chance for rehabilitation. "The children who keep the trauma deep inside have the most trouble settling back into normal life," Jacqueline tells me.

"But when you've done such shocking things, even under compulsion, is it possible that you can ever live a normal life again?"

Jacqueline shakes her head. "No, it's not possible, they'll always be under suspicion from their fellow villagers, and they'll always be troubled by memories of the horror. They can never forget that, but we try our best to help them as much as we can. When the army brings us children they've just rescued, we give them paper and pencils and ask them to draw something."

She shows me childish drawings of villages under attack by the LRA, of people being hacked to death, of forced marches through the jungle with children carrying bundles on their heads, and one drawing that depicts a dismembered child on the ground and a big pot on a fire. LRA soldiers and children surround the pot and the mutilated body. "They do these drawings because their minds are still suffering from life in the bush. Some were even forced to kill and eat friends they grew up with in their village."

Jacqueline shows me more drawings, this time of the rehabilitation center. "The children do these after a week or two here, and that tells us the child's mind has traveled back from the bush, from the suffering, and that's a sign of recovery."

The last group of drawings shows a church, the child's home village, the former school, boys playing football. "That tells us the child's mind has traveled back to where he or she was before being abducted, and that means the child is ready to go home."

Tragically, some mothers refuse to take back their rescued children, and, more commonly, the child will never be accepted by the people from his or her village. "People say that all these children who've been with the LRA are rebels who have killed, and ask, 'How can we trust them not to kill again?' So, some of the girls hang around army units because they want to become wives of soldiers. The soldiers use them and then discard them,

A drawing by an LRA girl at the Children of War Rehabilitation Centre, the drawing made just three days after her capture by the army. It shows the LRA attacking her village and killing people.

and so they become prostitutes. And many return to their home village but can't settle down, and [ultimately] return to towns like Gulu where there are not many jobs."

The gates open, a military jeep enters the compound and a white-haired man with a sprayed on smile gets out. A soldier trails him brandishing an assault rifle. The children edge away from the older man as he walks among them, trying to start conversations. He reaches down to hug a tiny girl, but she escapes into the arms of a nearby woman, perhaps her mother.

"It's Banya. He was the third-ranking leader in the LRA for more than a decade, the brains of the rebellion, but he was captured by the army a few years ago," Jacqueline says, spitting out the words. "The children are terrified of him, but we can't stop him coming here because he's under the army's protection. He's a known pedophile, we've heard many stories from children that he raped girls as young as ten before passing them on

The LRA's former number-three commander, Kenneth Banya, under twenty-four hour guard at the army barracks in Gulu. Banya was described as the brains of the LRA rebellion before he was captured by the Ugandan army. He is allowed to move around Gulu with an army escort and is kept at the barracks.

to younger commanders. The children know this, and fear him. When they see Banya, it reminds them of all the terrible things that happened while they were captives of the LRA."

"Then why does he come here?"

"You can see he's held captive by the army. So, he's a much diminished man since they got him, and I believe he's nostalgic for the power he once had over all the LRA, including the children."

"But that doesn't explain why the army lets him come here, frightening the children."

"He makes propaganda broadcasts for them on a local radio station aimed at the LRA, and for that the army allows him to move around town, though he always has an armed soldier guarding him."

"Can I talk to him now?"

Jacqueline shakes her head. "You have to ask permission from the colonel in charge of the 4th Division here in Gulu."

She leads me to the rear of the camp where, beneath one of the handful of trees in the dusty compound, twenty-year-old Rosemary Livingstone, a child-mother, cradles her four-year-old son in her lap. Jacqueline tells me that she was one of the many wives of the LRA logistics chief, Lieutenant Colonel Opira Livingstone, who was killed in battle a year earlier.

Kony's commanders enjoy multiple marriages to girls abducted from their villages: the higher the rank, the more the wives. "Kony has fifty-two known wives and twenty-five little girls presently tucked away ready to become his newest wives when they reach puberty," Jacqueline says, knowing this from hundreds of debriefing sessions with escaped women and children. "Kony wants the daughters and sons of the child-mothers and his fighters to form the core of his new nation. His wives keep a check on the girls set apart for him, informing him when they begin to menstruate, the time when he marries them. He's very afraid of contracting AIDS, and so takes only virgins as his wives."

Rosemary is a handsome young woman, tall, full-lipped and charcoal-black like most Acholis, but her eyes register no emotion, none at all, as if they have been stunned permanently by the terrible things she has witnessed. She was abducted from her village, about twenty miles from Gulu, when she was eleven. "About thirty LRA soldiers arrived soon after midnight and shouted that we all had to come outside our huts. They bayoneted my father and mother to death in front of my eyes, as well as most of the other adults, and then forced all us children to go with them."

Was she ever forced to eat human flesh? I wince as I ask the question, not wanting to stir traumatic memories, and yet unable to think of any other way to phrase it. Rosemary keeps silent for a few moments, her throat pulsing, and then nods.

"The LRA troops robbed our village of all its food, and made the twenty or so children they abducted that night carry the food in baskets on our heads. To escape the army they marched us through the toughest jungle, and after a few hours many of us were very tired and wanted to stop and sleep. The soldiers threatened to kill us if we didn't keep moving. I think they feared that the army, on learning of our fate, would chase after us. As morning neared, a little girl I knew well, Mary, she was seven years old, fell to the ground and began to moan. A soldier kicked her in the stomach, attempting to force her to stand. She moaned louder. He threatened to kill her if she didn't get up, but she closed her eyes and stayed on the ground."

Rosemary utters a deep sigh. "What happened next I wish was just a nightmare, that I dreamed it, but it's true. The soldiers ordered us to beat Mary to death with tree branches. We hesitated, but they threatened to shoot us if we didn't. We began to hit her hard on the body and head, and she was so small that she quickly died. Two of the soldiers then chopped her into pieces with machetes, while other soldiers took our branches and set them alight under a cooking pot. They put into the pot the pieces of Mary and her blood, which they'd made some children collect in bowls, and when it was cooked they made us each eat the flesh. I felt sick in my throat, but we had no choice. They threatened to kill us if we didn't. They told us that if any of us tried to escape, we'd suffer the same fate as Mary."

I see from the corner of my eye that Jane has tears streaming down her face. My own eyes glisten. I have seen more bloodshed covering stories than I wish to remember, and terrible examples of human cruelty, but nothing comes near this atrocity. If it were just the testimony of Rosemary, then I would remain skeptical, because she might be making it up as a way of revenging herself against the LRA. But Jacqueline says the killing and subsequent forced cannibalism is a common experience for many of the children whom she and the other counselors debriefed at the center. "We have files piled high of our counseling children, and many, oh so many, tell similar stories even though they were abducted in other places many, many kilometers apart, and were force-marched to LRA camps."

Rosemary suffered the torture more than once. "A couple more times on the march which took us several weeks to reach the LRA main base just inside southern Sudan. Each time we had to kill and eat a child who tried to escape. In Sudan, more children tried to escape, and when the soldiers caught them, we had to kill and eat them."

On the march she saw LRA troops murder villagers that they met on the paths through the jungle. "They usually beat them to death. Even though I was just a girl I knew that I couldn't show I was frightened because they warned us that they'd kill us if we did, and so I tried to hide my feelings, but I felt very bad about it."

Rosemary is Banya's niece, but she says that did not stop him raping her when she arrived in southern Sudan. Banya still wanders around the center, and I notice that Rosemary has faced away from him. "She still fears him," Jacqueline explains.

The first tear appears in Rosemary's eye and she hugs her son tighter to her breast. "I was unwilling to have sex with him [Banya] because I was just eleven, and it should never have happened even if I was old enough because we were from the same family. But he put a gun at my head. I trembled with fear. 'Do you know how many people I've killed?' he said. 'I'll shoot you if you don't have sex with me.' I did what he wanted. He kept me for a month until he found another little girl. Then, he gave me to Opira to marry."

"Did you end up loving your husband?"

"No!"

"But were you sorry when he was killed in battle?"

"Yes, because he was the father of my son."

"What's his name?"

"The same as his father's, Opira."

"What will you tell him when he's old enough to understand?"

"That his father was a rebel, but he looked after us well, giving us shelter and enough food."

Rosemary wipes away the single tear from her cheek and rocks her little son from side to side until he falls asleep. Jacqueline takes me to talk to several other child-mothers who all tell much the same story, of being abducted, prodded at gunpoint into forced marches to LRA camps, and being forced to kill and eat other children who tried to escape or who fell exhausted in their tracks.

The most chilling testimony comes from a stick-thin nine-year-old, Julia Okello, who had been rescued by the army in a raid just five days earlier after three years with the LRA. Her legs and arms carry the scars of what she tells me were many beatings while the LRA held her captive. We sit in the shade and, unlike many of the other children who avert their eyes when talking about their experiences, Julia looks me straight in the eye as she tells of the night the LRA soldiers attacked her village, just ten miles from Gulu.

"I was six years old when the LRA soldiers came for us. It was late at night and I was asleep, but woke when I heard banging on the door," she tells me in a steady voice without a hint of emotion. "As my father went out a soldier hit him on the head with the bottom of his rifle, then stabbed him to death with the knife attached to it. My mother screamed and another soldier chopped her across the neck with his machete. I saw the blood pour from the wound.

"There were about forty soldiers, armed with rifles, and they dragged all the children out of the huts and made us form a single line and roped us together. We all had to carry things they'd stolen from our village. I had a can of oil strapped to my back and though it was heavy, I was too frightened to complain. I didn't look at my sister or my playmates, just stared straight ahead as the soldiers made us march into the jungle. As we left, we could hear gunshots and I think they were killing our parents.

"We walked for many days, sleeping in the forest and using small tracks during the day. About a week after we were taken, we came across a man riding a bicycle along a small path through the forest. It was a Friday. I didn't know it then, but under Kony's Commandments, riding a bicycle on Friday was a serious sin, and the punishment was death. The soldiers gave six of us children machetes and forced us to chop the man to death. They threatened to kill us if we didn't do what they said. I closed my eyes and took the machete, though I felt so faint that I could barely grip it.

"None of us wanted to be the first to chop the man. A soldier hit one of the girls in the mouth with his rifle. She fell with blood coming from her mouth. The soldiers shouted at us to kill the man quickly. I turned my eyes away from the man who was screaming and hit him a few times. He died quickly.

"Two of the soldiers chopped him into pieces and put them into two big iron cooking pots we were carrying, along with the man's blood. They then made us chop up several papayas we'd collected earlier in the day and put them in the pots. We had to stir what was inside as the soldiers made fires under the pots.

"I didn't believe the soldiers at first when they said we'd have to eat the man's cooked flesh. I bit my lip and shook my head, but when I glanced at my sister, her eyes told me I should not refuse. The soldiers told us that if anyone didn't eat the man's flesh then they would kill that child and make the other children eat the flesh. I was given two pieces in a bowl with the soup made from the blood. I tried to think of nothing, to make my mind go blank as I ate the man's flesh. I can't remember what it tasted like.

"It took several weeks for us to walk to Sudan where Kony had his base and during that time we had to kill and eat one of my playmates. He was too small to keep up with us, and when he refused to go any farther because he was too tired, the soldiers made us kill and eat him. We had to kill by biting him to death."

Jacqueline leads us back to the camp's entrance. "Julia's experiences are not unusual for these children, even though she was just six at the time. We have so many files where children just rescued tell us much the same thing. Their testimony has to be true because it's not possible all these hundreds of children could make up the same tale with much the same detail, especially when you consider that the army has been rescuing them from camps all over northern Uganda and even in southern Sudan."

"But why would the LRA do this? What did their leader Kony have to gain from forcing children to become cannibals?"

"Who knows what lurks in the mind of that evil man, but from talking with all these children, I sense he wanted them to be so consumed by the guilt of what they did that they could never go back to their families. The shame would be too great, and would haunt them for the remainder of their lives. Also, their experiences would make those villagers who survived the LRA attacks be hostile toward them if they came home, knowing they beat and killed other people, even eating them.

They'd consider them so tainted with the LRA's evil ways that they were beyond redemption."

"From your experience, are they?"

Jacqueline sighed. "It's too early to know. They should be counseled by experts for years to come, but we don't have the funds, so they get at the most two or three weeks of counseling here, and are then taken back to their villages. We tell them that because they were forced to kill even their friends, forced to eat human flesh, that they should not feel guilt. They had to do it or be killed and maybe eaten themselves. But, realistically, the trauma will stay with them all their lives. Sometimes they find that their parents are alive, but often they were killed in the LRA attacks, and some of the children even had to kill their own parents and help burn the village huts. So, the children must live with relatives who may not be sympathetic to their trauma. Every rescued girl comes home pregnant, or raped, or bringing back two or three children."

The repeated recitation by children of forced monstrous actions almost beyond imagination has begun to numb my mind. Jane and I walk back to the Acholi Inn in silence. We go to our separate rooms, and I put my head on the pillow, hoping through sleep to escape the turmoil that is shaking my mind. But the images evoked by little Julia and the others at the Children of War Rehabilitation Centre will not go away, will not be denied their opportunity to torment me with their horror.

These tales of mass forced cannibalism have been extensively documented by other international agencies in Gulu. Among many others, the United Nations Office for the Coordination of Humanitarian Affairs documented testimony from a fourteen-year-old boy identified as O.R. He was abducted by the LRA from his home in the adjacent Acholi province of Kitgum. "On the way to Sudan, they forced us to kill many people. One morning, a young boy was brought to us. We were told he had tried to escape. His body was swollen and had cuts from many beatings. They killed him. We were told to chop the body into smaller pieces. Boys were given the heart and liver to eat. Girls were told to cook and eat the rest of the body parts. We did as we were told."

The UN reports that O.R. was also forced to pound a baby to death with a wooden mortar. He was rescued by the army, but is still tormented by his actions. "I am constantly disturbed by what I did in the bush. I dream about it all the time. Sometimes I hear voices saying things to me. 'There is

work waiting for you in the bush,' the voices keep telling me. 'Pound faster
. . . harder . . . harder,' other voices keep saying. In the night I dream of
the same things. I fear to go to sleep because of nightmares. I want these
dreams to stop tormenting me."

That same torment is the catalyst that night for one of the most troubling
sights I have ever experienced in four decades of reporting from remote
and often dangerous places around the world.

CHAPTER 18

As the light fades from the northern Ugandan sky, Jane and I drive from Gulu's town center along a road leading to Sudan. The bitumen gives out after three miles, the road transformed into a dirt track pressed on both sides by mud hut villages. At the seven-mile marker, a place called Lacor, we pass St Mary's Hospital, one of the largest in the north. "That's where we had the outbreak of Ebola," the driver says.

As if the inhabitants of Gulu are not already tormented to the limit of human endurance by the LRA scourge, it was also an epicenter of one of the world's most horrific outbreaks of the disease. In October 2000, sixty-two patients turned up at Lacor with symptoms of the then little-known virus, Ebola, suffering from headache, vomiting, loss of appetite, diarrhea, severe fatigue, abdominal pain, body aches and copious bleeding from all the orifices. Scientists believed Ebola, also known as viral hemorrhagic fever, was first picked up by a human from an infected great ape, perhaps a gorilla.

The virus is transmitted by infected body fluids. Gorillas and chimpanzees share up to 98 percent of our DNA, and so disease can pass readily between the great apes and humans. Jane, who covered the story, says that "it's possible someone handling raw bloody meat from a gorilla infected with Ebola picked up the virus and then passed it to family members caring for that person when the fever struck. It spread quickly to communities in Uganda and the Congo, which is not far from here."

By December, Lacor had 425 patients desperately ill in quarantine wards, and the hospital's doctors and nurses battled valiantly to contain a virus the world then knew little about, a disease that had no cure. "The hospital's medical superintendent, Dr Matthew Lukwiya, and twelve nurses were infected with Ebola and died," she says. This tale of true heroes risking their lives to help others in need braces me, and for a while nudges aside my brooding over the demonic nature of the LRA rebels.

A mile beyond the hospital, the driver pulls to the side of the road. We are near the equator and the light goes out at night as suddenly as a switch turning it off. The road is mostly deserted with just a few men on bicycles peddling slowly home or toward Gulu. "We won't have long to wait; the children will come soon," the driver says.

Just after seven I see the first silhouettes, like ghosts, coming over the hill and heading toward us. They are children who have left their families' mud huts to make the long walk along dirt roads to Gulu, the nearest town. We remain in the car as the first wave of what will amount to thousands of children pass by this spot tonight on their way to sanctuaries in the town. They grip sleeping mats as they silently trudge along both sides

At Gulu in northern Uganda, toward dusk thousands of children, the "night commuters," stream into town from their villages to sleep securely, thwarting any effort by LRA soldiers to abduct them.

of the road. Fearful-eyed toddlers hold their big brothers' hands or ride on their sisters' backs, while girls who look as young as eight or nine nervously peer at the roadside shadows as they pass. They are on the move because they live in a world where a child's worst fears come true, where armed men really do come in the darkness to steal children, and their shambling daily trek to safety has become so routine there is a name for them: "night commuters."

I leave the car with Jane and begin walking with one of the boys, his thin body wrapped in a patched blanket. He is ten years old and says his name is Michael Ochora. "Like all the children, he comes into the town every night from his village to evade the LRA," Jane tells me. Michael grimaces with fear. "The LRA attacked our village last year while I was staying with an aunty for the night at another village," he tells me as we walk. It is his choice not to stop. "They took away many boys and girls, my friends, and they've never come back. I can't get to sleep at home at night because I fear the LRA will come and take me."

For more than two hours the silent trudge of the children leads them into town where about twenty thousand bed down in special sanctuaries and schools with the overflow sprawling in alleyways, on shop verandas and at the bus station. Wrapped in their sleeping mats from head to toe, and fast asleep, they look like corpses laid out all over town at a disaster scene. Another twenty thousand kids flock nightly to the town center at Kitgum, an Acholi town about thirty miles from the Sudanese border.

As I look at the sleeping children, I remember the words of Jan Egeland, the UN's Under-Secretary-General for Humanitarian Affairs, when he came to Gulu. He called the situation the world's largest neglected humanitarian emergency. "Where else in the world have there been twenty thousand kidnapped children?" he asked. "Where else in the world have 90 percent of the population in large districts been displaced? Where else in the world do children make up 80 percent of the terrorist insurgency movement?"

Noah's Ark, on Gulu's outskirts, is one of the town sanctuaries for the night commuters. All night, a pair of nervous soldiers armed with AK-47 assault rifles stand guard over the sanctuary, a huddle of large tents in a field surrounded by a high barbed wire fence. About two thousand children have gathered here for a night of safety, to enjoy a free meal and a wash with clean water from a pump. Afterward there is an exuberant display of

The "commuting" children bed down in any available space, leaving their villages to spend the night securely in Gulu.

traditional dancing with the boys and the girls forming separate circles and swinging their hips and stamping their feet to the thud of drums played with captivating skill by other children.

Austin Ojara, a counselor for the boys at Noah's Ark, says that up to six thousand children come to just this one sanctuary each night. "Around 65 percent of them were abducted by the LRA and then rescued by the army, and so when they return home they fear to sleep at night because they could be abducted again. They know they are safe here. All have suffered terribly, but the children who had to kill and eat other children are the most traumatized."

Inside one of the tents, children have rolled out their woven-straw sleeping mats and sit on them as they do their homework, read books or chat. Jane takes me to George Kilama, a fourteen-year-old who spent three years with the rebels after being abducted from his village near Gulu. Slim, small and barefoot, he is clad in frayed shorts and T-shirt, and has a street urchin look that belies what he used to do.

"I was a soldier with the LRA for a year and fought against the army," he says hesitantly, perhaps unsure whether the *muzungu*, or white man, seated by him might report him to the government. "I was taught to use a rifle and killed government troops. In the last fight we ambushed the soldiers and killed eighteen of them. We took their weapons, but left their bodies in the open. But then the army attacked our camp with helicopter gunships and I was captured."

He was lucky to survive. An army spokesman, Major Shaban Bantariva, stated that the army was unrepentant about using helicopter gunships to terrify the rebels by shooting on sight LRA men, women and children, even toddlers and babes in arms. "The LRA train their women and children to use rifles and even rocket-propelled grenades, and so we shoot them before they shoot us," he said. He confirmed that official battle statistics list anyone from the LRA killed in an attack as an enemy, even babies.

The one foreign reporter to see the aftermath of a helicopter gunship attack on an LRA camp was Callum Macrae of the BBC, who was taken to such a scene a day after it happened in March 2004 near the Sudanese border.

"It was a scene of terrible carnage," he reported. "Dozens of bodies lay scattered around the undergrowth where they had fallen. The first body I saw, the first of these fifty-five rebels, was about four years old. Some ten meters away a girl lay dead, stripped to the waist. She may have been the child's mother. One of the [dead] women looked as though she were still alive, until I walked round and from the other side I could see the top two-thirds of her head had gone, blown away by rockets from one of the Ugandan army's new helicopter gunships.

"We were moving quickly through the bodies—the area was not entirely safe—when I heard a soldier say, 'This one's alive.' He was a boy, of fighting age certainly, so perhaps fourteen. He was lying semi-conscious, his chest shuddering. He had lain there, unattended, for nearly twenty-four hours.

"'Can't we get medical attention for him?' I asked. 'We will carry him back and treat him,' I was told. But then five minutes later a soldier brought back the news he was dead. This is a truly awful war."

George lived through one such strike. "Two helicopters attacked without warning as we were eating our midday meal by a stream," he tells me. "I huddled against a woman on the ground and a rocket smashed into her side splattering me with blood, killing her but not harming me. Maybe that's why they didn't shoot me from the air, because they thought I was dead. When they landed they took those of us still alive back to Gulu."

However traumatic the attack was on his young mind, what George suffered three years earlier at the hands of the LRA was far worse. It is why each night he flees from what was once the safety of his village, and his family, and trudges into this sanctuary to stay until dawn because he knows the LRA usually attack villages only at night.

I brace myself as Jane patiently listens to his testimony, takes notes, and then translates for me. "I had to take part in killing many children who tried to escape or who could not keep up with us as we marched to the Sudan. Then, we had to eat their flesh. Once the soldiers ordered us to tie to a tree a little girl who slept too long each night, set her alight and watch as she burned to death. I was so sad, she was a nice little girl, but I dared not complain or the commander would have killed me."

His commander made George and other youngsters slaughter a seven-year-old boy so they could collect his blood in saucepans and warm it on the fire. Then, the commander ordered them to drink the blood, threatening to kill any who refused. "Drinking the blood strengthens the heart," George says, repeating what the commander told him. "You then don't fear blood when you kill a soldier or see someone dying."

As with Julia's group, children were slaughtered because they were too tired to walk as the rebels marched to another camp through heavy jungle or fled an army attack. One night a pair of five-year-old boys fought by the fire over some trivial matter, and were then too tired to walk when the group broke camp. "The commander ordered us to kill them with *pangas* [machetes], but this time we didn't have to eat them as the commander was in a hurry to leave."

George suffers from nightmares and has trouble sleeping. "I don't ever want to see a gun again," he says. "Never, ever, again."

The boy seated beside George, a skinny twelve year old named Opira Nelson, stares at the mat as he describes helping to beat a boy to death with logs because he tried to escape. "The commander ordered us to lick the blood from his body," he remembers. "He said it would make us brave when we became soldiers and fought the army."

The horror deepens. Robert Ochira is now fourteen and has returned to school, having lost two years after he was abducted from his village by LRA troops at night and made to be a porter while he trained to be a child soldier. He was forced with other children to chop into small pieces the body of an escapee child they had just killed with machetes. "We then had to cook and eat the flesh," he tells me. "We did as we were told because we didn't want to die like that."

A few days later, Robert's commander ordered him to pound to death the newborn baby of a woman who had been captured a few nights earlier, just like O.R. "The baby's crying was annoying the commander. I was given a large wooden pestle, and the commander told me to keep pounding the baby's head until it was dead. We then had to cut it up and eat it."

As if to legitimize the terror with an overlay of mystical and religious ritual, the LRA leader, Joseph Kony, mingles traditional Catholicism with witchcraft and the cult of Lakwena, the murderous spirit first encountered by Alice. Kony's time as an altar boy had steeped him in the ancient rituals of the Catholic Church, and from the buzz in his mind he formed a uniquely twisted version of Christianity to induce fear and obedience in his "flock." Child after child at the shelter tells me of the bizarre rituals practiced once they arrived at their first LRA camp, and when their soldiers were about to go into battle.

"Jane," a nom de plume and not my translator, had told the counselors at Gulu's Children of War Rehabilitation Centre that she had been held captive by the LRA for eleven years, and described the rituals: "When a recruit arrives he or she is taken to the river by the controller [the unit's commander and also spiritual leader]. For me, he prayed to God to purify me with power of water and camouflage. He poured water on my head, mixed with camouflage. For three days a recruit is initiated by smearing shea nut oil on the forehead, heart and back. He or she can then eat with people immediately. I had to stay for four days without eating with others. After that I become a full member.

"We had monthly purification rituals. Everybody goes to the river, but male and female go separately. This is ordered by the spirit and is done using river water very early in the morning. When going for battle, all fighters are taken to the field. The controller goes ahead, [the fighters] bare-chested with their guns. The controller will sprinkle water and purify their path. They sing and water is sprinkled on them using Olwedo leaf—symbolic of goodwill and blessings. Controllers pray for them and bless them."

Kony told the captive children, and those who had been brainwashed into following him, that along with Lakwena, the most powerful spirit, he was given instructions by Dr Ambrosoli, the ghost of an Italian doctor who had worked in an Acholi hospital during the World War II. Surrounding

Kony in a protective huddle were ten more powerful spirits including three Americans, one improbably named King Bruce after kung fu star Bruce Lee, and two Chinese.

The mumbo jumbo worked. "Kony always knew from Lakwena and the other spirits when the Ugandan army was near," Robert tells me. "That's why we soldiers usually won the battles, because we could set up ambushes. But if the spirits told Kony that there were too many Ugandan soldiers coming, then we always had time to flee to safety."

A boy named John Okello tells me that the LRA move easily through the bush near Gulu every night. "Today I went to my grandmother's village for her burial. It's about five miles from Gulu. My relatives told me that the LRA raided a village nearby last night and took away children. So, it's still very dangerous for children to stay in their villages at night, and that's why we come here, because we don't want to be abducted by the LRA again."

Once again we return to the hotel shocked into silence. Jane had heard such tales on a previous visit to Gulu, but they still caused her immense sadness, the horror stilling her usual bubbly self, and her face, usually so bright and cheerful, was now suffused with grief. I thought I was toughened to the terror and brutality of war, but I was not prepared for meeting these children. That night, surprisingly, I fall asleep almost as soon as I lay down on the bed, perhaps to escape from my own mind. But the nightmares come soon enough, and I wake several times during the night, my head buzzing with visions from the children's stories. Sometimes I am the victim killed and eaten, and sometimes the murderer.

CHAPTER 19

The sky is still a chalky black when I wake, again fleeing a dream where I was a prisoner of the LRA who forced me to do the monstrous things Kony has thrust upon the children. The power supply has crashed, a daily happening in Museveni's Uganda, and so I use a torch until sunrise to read a book I found in Kampala. *Aboke Girls*, by Els De Temmerman, published in 1995, tells of one of the first LRA mass abductions of children, in October 1996. Using midnight darkness as cover, the rebels seized 139 schoolgirls at gunpoint from their dormitory at St Mary's College, one of the northern province's best schools, and forced them to march for several weeks over rough terrain to Kony's base just inside southern Sudan.

The entire nation was outraged, and President Museveni ordered the army to get the girls back. They failed, unable then to cross the border in pursuit, but Sister Rachele Fassera, a brave Italian nun and the school's deputy headmistress, followed the LRA troops all the way to Sudan and pleaded for the girls' return. Kony gave her one hundred girls to take home, but he farmed out the remaining thirty-nine among his commanders as minor wives.

☠

At 8 am I leave for Gulu's Catholic cathedral, about ten miles from town, the bumpy muddy dirt track skirting the 4th Division's headquarters. I will be entering there soon enough, granted an interview with the commanding colonel later in the morning. Spread across a hill and guarded by barbed wire and armed sentries are scores of small African mud huts with conical thatched roofs, the married soldiers' quarters.

Driving me to the cathedral is an Anglican bishop, Maclord Ochima, a grizzled man with grief etched forever into his bluff features. I wonder that he can keep his sanity when he tells me that his wife was killed by an LRA landmine while driving along a road near Gulu, and his daughter was raped by the rebels. She then committed suicide. Along with the Catholic Archbishop, he is a leader of the ARLPI, or Acholi Religious Leaders' Peace Initiative.

He and the Catholic Archbishop at Gulu had met with senior LRA leaders and were negotiating an amnesty that promised to end the conflict when Museveni called in the International Criminal Court, based in The Hague, which issued warrants for the arrests of the senior LRA leaders including Kony. The warrants alleged that they "engaged in a cycle of violence (in northern Uganda) and established a pattern of brutalisation of civilians by acts including murder, abduction, sexual enslavement and mutilation, as well as mass burnings of houses and camp settlements."

Bishop Ochima scorns the warrants: "We Acholi leaders condemn the action of the ICC because its involvement will destroy any hope of persuading Kony and the other LRA leaders to give up their weapons and make peace," he tells me. "It's wrong for the ICC just to investigate the crimes against humanity of Kony and his cronies. They should investigate all the crimes against the Acholi people, including the many murders committed by the army. The army has even buried people alive. There are many eyewitnesses to those atrocities."

Archbishop John Baptist Odama, a stocky man with a genial smile, ushers me into his study in a walled compound just beyond the cathedral. He is clad in a spotless white soutane, and a gold crucifix dangles from his neck. The Catholic prelate is head of the ARLPI, and spends hours each day attempting to end the LRA rebellion and bring all the abductees home.

"Twenty years is too much, this is a full generation that has known nothing but war and fear, and so the ICC should abandon their quest to arrest Kony and other LRA leaders," he tells me. "Children born during the war are adolescents—some have children, and parents have grandchildren. It is the Acholi way to offer amnesty to even your worst enemy. We have traditional ceremonies that bring together combatants and heal the wounds, and a full amnesty is the only way we will end this terrible conflict. But while the ICC is involved, Kony will never trust any amnesty offer."

"But isn't that like offering an amnesty against his war crimes to Adolf Hitler if he were alive today?" I ask. "The ICC claims the crimes committed against humanity by LRA leaders are so terrible, so beyond the pale, that they can't comply with your request."

"That's easy to say when you live far away from the danger in Europe. Their children don't go to bed every night fearing they'll be dragged from their sleep and forced to join the LRA. Our children are still being abducted by the LRA, and 90 percent of Acholis have fled to government camps where the army soldiers rape the girls and torture the men. They are kept at the edge of starvation by insufficient rations. Which is the greater evil? The ICC should go back to Europe and leave we Acholis to solve the LRA crisis."

The Archbishop has spoken to Kony many times, in person at one of his bush camps and by cell phone, and says he comes across as an intelligent man who has carefully thought out his ambitions as a rebel leader and how to accomplish them. "Kony makes the children kill and even eat each other so they feel such an enormous sense of shame and guilt that they can never go back their families and their villages," he says, echoing Jacqueline Okongo's belief. "The only family they then have left is the LRA with Kony as their father. It's part of his plan for a racially pure Acholi nation led by him."

There is a knock on the door. In walks a middle-aged European woman clad in a loose skirt, sensible shoes and an abbreviated nun's coif that covers her cropped hair and flows down her back. Her lips are drawn tight, and her eyes, behind a pair of thick-rimmed glasses, are suffused with a haunting sadness. I sense that she does not find much to smile about each day, perhaps after suffering an overwhelming loss.

"Sister Rachele, welcome," the Archbishop cries with genuine warmth and holds out his hands to her. He turns to me. "I'm sure you've heard of Sister. She was the hero who brought home most of the Aboke girls."

I ask her whether, a decade after their abduction, the girls have all been rescued. "Most have," she answers in accented English, "but some girls are still out there held captive by Kony and the LRA. We've not been able to get them back. God have mercy on them."

Sister Rachele has come to the Archbishop to report an atrocity committed by the army about twenty miles from Gulu. "The soldiers killed and buried four boys they believed were LRA members," she says. "We have reputable witnesses."

☠

Bishop Ochoma has no wish to enter the army compound today and decides to stay for a chat with Archbishop Odama. So, Jane and I take a taxi to the 4th Division headquarters. We are driven through the gates, guarded by armed sentries, and after a short wait are ushered into the office of Lieutenant Colonel Nathan Mugisha, the division commander, who doesn't rise or even look at us as we enter. He continues to study a file he is reading for several minutes, leaving me in limbo. I know his name and rank because he has one of those kitschy slabs of wood carved with these details on his desk which fronts a meeting table that fills much of the room. Scattered across it are well-thumbed files. I ignore his bad manners, his amateurish attempt to show me who is the boss.

Mugisha puts down the file and turns his hard eyes on us. He looks a trim spit-and-polish martinet, and has a no-nonsense stare that threatens to cut right through me like a laser. In high school I had a teacher like that, and we feared his piercing stare far more than the cane-wielding teachers. They were just one-dimensional brutes, but the starer seemed somehow demonic.

Museveni has ordered Mugisha to smash the LRA rebellion, but he is not making a very good job of it. Yet, he is especially dismissive of the LRA child soldiers even though they have often beaten his own troops in firefights. "They're illiterates and they're young, they're infants who can easily be moulded to whatever Kony wants," he says. "He looks at that innocent stage, at that fragile stage where you can easily brainwash the children, and that's why he doesn't go for mature people, that's why he hates people who've gone to school. He himself never finished primary school."

And yet Kony has consistently kept an army of twenty thousand troops on the run here in the north. Mugisha waves a hand across his face, dismissing my sally. "We've got *him* on the run, and we'll get him and the other leaders pretty soon."

"The army's been saying that almost since Kony began fighting you two decades ago, and yet you don't seem any closer to defeating him."

Rather than being offended, the colonel seems energized by my question. "Give us another year and then come back. Kony and the rest of them will be in jail, I promise you that, and we'll never let them out."

I ask Mugisha about Sister Rachel's atrocity claim. "Ha! That woman is always attacking us. Let her bring me the proof and then I might take her seriously. She achieved wonders in getting most of her girls back, and now she should go back into the convent and stay there."

I ask if I can meet Kenneth Banya, once the third-ranking leader of the LRA.

"Of course, we've got nothing to hide. We're holding him under guard in a bungalow just down the road. You can go and meet him now if you want."

Banya was captured in July 2004 after a fierce battle in Gulu district. One of his wives and four-year-old son were killed by helicopter gunship fire, but most of his 135 soldiers got away. I was told in Kampala that Banya was tracked through the bush using satellite technology secretly funded by the US. When I asked the US Ambassador in the Ugandan capital whether this was true, he replied with a smile: "If it were, I wouldn't tell you."

The seventy-year-old Banya is kept under twenty-four-hour guard, living with other captured LRA officers at the Gulu barracks, but wanders around the town accompanied by a soldier wielding an assault rifle. The army uses him for propaganda, with Banya constantly broadcasting over a Gulu radio station its call for the rebels to surrender.

When I meet him at the barracks he denies with a sleazy smile that he had ordered children to be murdered or had sex with young girls, and claims he was an unwilling rebel, abducted by LRA fighters in 1987. Banya was a military helicopter pilot, trained in Moscow and Texas, and Kony welcomed his prowess in military tactics. Lieutenant Paddy Ankunda, the northern army command's spokesman, laughs when I tell him Banya's tale. "He crossed over to Kony when the present government seized buildings in Kampala he'd grabbed in an earlier coup."

Banya was Kony's senior advisor and strategist and knows him better than anyone. He believes Kony wields supernatural powers, using an enigmatic and apocalyptic spiritualism to control his people's minds. Kony has rarely been seen by the outside world since he went into the bush two decades ago, but a picture shows him looking like a crazed rapper in a cowboy hat, dreadlocks and sunglasses. Banya says he claims to be God's instrument on Earth and uses obscure Biblical references, gleaned from the Old Testament, to justify his lust for extreme violence including count-less camp executions, burning enemy villages and hacking off enemies' body parts. His catechists lead daily prayer meetings at which they preach his perverted form of Christianity mingled with Islamic tenets and tradi-tional Acholi beliefs. Banya adds that they pray with rosary beads, and that Kony forbids the eating of pig meat, a crime he has decreed punishable by execution.

He tells me much of what "Jane" had related, describing the purification rituals administered by the controllers. Fighters were sprinkled with holy river water before battle for protection from bullets, and a new initiate—such as Banya when he arrived at his first LRA camp in 1987—was sprin-kled with water on his bare torso before being marked with crosses made from white clay mixed with nut oil. "That removes your sins, you're now a new person, and the Holy Spirit will look after you," Banya tells me.

This all-knowing Holy Spirit, the same Lakwena who possessed Alice, guides Kony's decisions, including battlefield tactics. In front of his people Kony talks openly with the invisible Lakwena, who seems to be an amal-gam of several ghostly identities, including an American bizarrely named, according to Banya, James Brickey Who Was He. "The Holy Spirit would inform Kony when the enemy was coming, or when people were doing bad things."

Despite Kony's reliance on an American ghost, Banya tells me that he regards Americans as "bad people." His animosity stems from the decade leading up to 2002, during which the LRA hunkered down in safe havens across the border in turbulent southern Sudan, protected by its tangled po-litical alliances. The devoutly Christian Museveni was then supporting the Sudan People's Liberation Army, a Christian-based rebel group fighting for independence from Sudan, battling its military rulers, the fundamen-talist National Islamic Front (NIF). In retaliation, the NIF supported the LRA in its fight against the Ugandan government. Banya says Kony helped anchor Sudan's valuable support by claiming America to be his enemy.

A grateful Khartoum rewarded him with anti-tank weapons, anti-aircraft guns, sub-machine guns, ammunition and food. Most importantly, Khar-toum gave the LRA land for villages near the southern Sudan town of Juba. There, safe from attack, rebels grew crops, seeded their baby farms, brain-washed and trained new abductees, and regrouped after strikes across the border. "We had seven thousand people there then," says Banya.

Thank Presidents Bill Clinton and George W. Bush for helping end this cosy affair. According to a well-informed embassy source in Kampala, the US government applied constant pressure on the Khartoum government to end support for the LRA. Success came in March 2002 when Khartoum signed a military protocol with Kampala allowing Ugandan troops to launch attacks on the LRA in southern Sudan. The Ugandan army, in a major cross-border offensive code-named Operation Iron Fist, swiftly destroyed the main LRA camps.

But forcing the LRA base camp back into Uganda enraged Kony and he ordered a destructive upsurge in fighting in the north along with a massive increase in abductions. According to UNICEF, more than ten thousand children were abducted between June 2002 and December 2003.

This vicious retaliation prompted Museveni to order an exodus of the Acholi population to the relative safety of government concentration camps, abandoning tens of thousands of square miles of some of Africa's most fertile land. "When we started our current assistance program in April 2002 there were 465,000 in the camps, displaced by the LRA," I am told by Ken Davies, director of the UN's World Food Program (WFP) in Uganda. "By the end of 2003 there were 1.6 million in the camps."

To visit one of the 139 camps in a WFP supply convoy, a line of trucks stacked with food and about half a mile long, is like going to war. Ongako camp is just ten miles from Gulu town, but one hundred heavily armed soldiers and two armoured vehicles mounted with machine guns accompany us to ward off any LRA attack. In my SUV, bullet-proof vests and helmets are placed on the seats for our use.

Ongako houses 10,820 gaunt internally displaced persons, or IDPs. In ragged clothing, like living scarecrows, they wait in long lines at a field nudging the perimeter of hundreds of small conical mud huts pushed against each other. A shiver of anticipated pleasure ripples through the desperately hungry crowd as WFP workers begin unloading the emergency rations that keep them alive: maize grains, edible oil, cooking oil and pulses, along with corn soya fortified with vitamins and minerals.

For the IDPs, life is a desperate daily struggle to survive. The World Health Organization says that a person in a sedentary environment needs 2,100 calories a day not be become malnourished. The WFP provides the

Children at Ongako. The IDPs were in a continual state of starvation because of lack of food at the camp. Many died of malnutrition. Here are some of the children eating food supplied by the UN. The boy on the left is suffering from severe malnutrition.

camp dwellers with up to 74 percent of that at an average cost of $31 a year per person. Much of it comes from USAID, with the camp dwellers expected to make up the difference with food grown in fields near each camp.

This is not easy. The government forbids camp dwellers from going further than a mile into the bush, fearing that collaborators will secretly grow food for the LRA. But just a football punt into the bush from a camp's perimeter, residents chance an encounter with a marauding LRA band, risking mutilation, death and abduction.

Camp leader John Omona squats by one of the huts and tells me that despite the WFP aid there is not enough food and medicine in Ongako, and it suffers from a drastic shortage of clean water. "The children get sick and die, and so do the adults," he says.

The Ugandan government provides no food for the IDP camps and the constant near-starvation is starkly obvious, especially among the children who make up 53 percent of camp residents, World Vision says the acute malnutrition rates are high, up to 21 percent. Hordes of kids wander the camps with the swollen bellies and red-tinged hair of kwashiorkor, the dreadful disease that stems from extreme malnutrition, and many die from starvation or from hunger-related diseases. A survey of the IDP camps by Médecins Sans Frontières showed that the daily death rate was at the level of "an emergency out of control," an emergency rarely glimpsed by the outside world. Darfur captures the headlines with its heart-searing pictures and stories of devastation and disaster, and there seems not enough room in the databases of most media to expose the nightmare in northern Uganda.

Dr Benjamin Abe, an anthropology professor in Seattle and prominent member of the Acholi diaspora, was horrified by his recent visit to an IDP camp near Gulu. "It was inhumane, basically a concentration camp," he tells me when we meet in a hotel lobby. In a meeting with President Museveni, Dr Abe says he urged the Ugandan president to close down the camps immediately and send the people home. Museveni refused. "His concern was that the LRA insurgency was not completely contained."

For an unsupervised encounter with the IDPs, the following day I take a car to another camp twelve miles from town armed only with advice from a local friend: "If the road is full of people you're OK, but if they suddenly disappear, then get back to Gulu as quickly as you can because it probably means the LRA are nearby."

Like Ongako, Awer camp nudges the roadside, a gigantic huddle of thousands of small conical family huts. The air is sour, fouled by the smell of unwashed bodies, sickness, and the stink that comes from years of fear, despair and hopelessness. With no jobs or fields to nurture, the men slouch in the shade of their huts or play endless games of cards. The children squat on the bare earth in mud-hut classes, up to ninety crammed into a room with no pencils or books. The women, weary from hunger, cook meager meals of maize or slowly sweep the dust from the family hearth.

About fifty men and women gather around me, victims of the army that guards them. The men's legs, arms and heads have been broken and scarred during torture while all the women claim they have been raped by soldiers. Grace Adoch, who is in her thirties but looks twenty years older from the grief etched into her bony features, tells me that a soldier accosted her on the road near the camp. "I was coming back from the hospital and the soldier called to me, 'You come here, you give me money.' I had no money, and so he took me into the bushes and raped me. When I tried to escape he said he'd shoot me."

The soldier has since died from AIDS, but Grace is too frightened to be tested. "It's very common for soldiers to rape women in the camp," she tells me.

UNICEF's Rob Hanawalt in Kampala later told me that little girls in the IDP camps fear going to the latrines at night because "they're afraid of being raped by soldiers, or even by other men in the camp."

The Awer camp leader, Oryma Francis, explains to me that long-time camp living has ripped asunder the Acholi tribal fabric, causing severe social problems. "The men get drunk and the bigger girls sell themselves to the soldiers to get money for their families. There are few marriages taking place. The rate of AIDS here is double that of the rest of Uganda."

☠

US legislators have long known about the LRA and its horrors. In July 1998, Jemera Rone of Human Rights Watch testified in Washington DC before the House Subcommittee on International Operations and Human Rights, and the Subcommittee on Africa. Rone revealed to the legislators a long litany of LRA atrocities including examples of child murders and abductions, but little came of it.

The world community has shown a similar lack of interest. Since 1999 there have been many UN Security Council resolutions demanding the protection of children in armed conflict, including calls for action against the LRA, but the pious words are never translated into action.

With camp dwellers facing catastrophic problems, the rescued LRA children still want to go home even to an IDP hut. In Gulu at the Children of War Rehabilitation Centre I witness many joyous and often stunned reunions between parents and abducted children. Ogwang Layet throws his hands in the air with unrestrained glee as he runs to his thirteen-year-old daughter Steler, seeing her for the first time in two years. "I thought she was dead, but I heard the army announce on the radio her name among the children rescued just two days ago," he tells me, his voice shaking and eyes misty. "I've not slept since we learned she'd returned. We're from Pader and we took a car here. The road is dangerous—there have been many LRA ambushes—but we were willing to risk anything to bring Steler home."

Her mother, Jerodina, is struck speechless by shock, but grabs Steler, pulls her head to her bosom and sobs uncontrollably. Just days after being captured by the army, Steler stares silently at the ground with haunted eyes, not responding to her parents, perhaps turned mute, like many returning children, by the shame of the atrocities she has been forced to commit.

The captured LRA children are from all over the Acholi region, and a few days later I fly in a UNICEF-chartered small plane from Gulu to Pader, a town thirty minutes away, with ten child-mothers and their babies returning home. They include sixteen-year-old Beatrice Ochora and her infant daughter. When she reached puberty Kony had farmed her out to a captain in his army.

"I was at Kony's headquarters for more than two years," she says over the drone of the plane's engines. "He's a very bad man. He sometimes gave us orders to kill a child who was lazy or who disobeyed him. We had to obey Kony or we'd be killed. We also had to kill children who tried to escape. We beat them to death with clubs or cut them up with axes. Sometimes we had to eat their flesh, but not always."

I turn away to look out the window and wonder what she can tell her daughter about life with the LRA when the child grows up and wants to know about her father. The terrain below would once have been sprinkled with hundreds of villages, schools and fields bursting with crops, but now with villagers having fled the LRA horror it resembles a land of ghosts, with not a hint of habitation.

At Pader an SUV takes Beatrice and her baby to a huddle of tents in a bare field, a holding center where LRA children wait to be returned home. As she arrives, a teenage girl rushes to her screaming with joy. "Beatrice, you're alive," she cries, and high-fives her. "We were best friends in the bush with the LRA," Beatrice explains. "She thought I'd been killed by the gunships."

The reunions bring happiness to the faces of most of the children, but they confront a grim future. There is almost no backup counseling for the fifteen thousand who have come home even though they have suffered far more trauma than almost anyone else on Earth. "They'll need counseling for years to recover from their terrible ordeal," says Jacqueline on my final visit to the Children of War Rehabilitation Centre, adding that there is little or no chance of that happening.

Jacqueline says parents are also marked for life. "When your child was

At the Children of War Rehabilitation Centre Beatrice Ochora attends a hymn-singing session with other former members of the LRA. The boys next to her had been soldiers in the LRA before capture by the Ugandan army.

abducted she was a small girl, but you're coming to meet someone who is now a woman with two or three children. Maybe your son was a very healthy boy, but now he's lost his leg, his eyes, his whole face burnt by bombs."

The villagers must also adjust to returning LRA children, many of whom are multiple murderers. Boys often resent the loss of the power they felt when they had an assault rifle, and girls find it difficult to slip back into village life. In a large tent crammed with boys who were until recently LRA soldiers skilled in shootouts with the army, a kid, about twelve years old, trains an imaginary gun on me. "Bang, Bang, Bang!" he shouts. "You're dead."

I retreat to the outside where I meet fourteen–year-old Peter Ochora, a stocky boy with tough eyes. Younger boys clear a path for him as he walks me around the center. His main ambition now is to return to primary school and catch up on the four years he has missed since LRA troops abducted him from his village near Gulu. His mother and father were mur-

At a primary school assembly in Gulu, the girls show none of the usual laughter African girls show when such a picture is taken. They had little to smile about: Many of them had been abducted by the LRA and returned home after rescue by the Ugandan army, and many were still under threat every night of being abducted by the LRA.

dered during the raid. But until two weeks ago, when the army captured him, he was willing to risk his own life to protect the LRA leader Joseph Kony.

"On my twelfth birthday in the bush I was given a rifle and trained as a soldier," he tells me. "I was so good at killing army troops that a year later I was made one of Kony's trusted bodyguards," a member of an elite force that guards the LRA leader around the clock. "Several bodyguards surround Kony's house, and another group stands guard at the camp's perimeter, and their task is to protect the leader. Then, other bodyguards are posted a few miles from the camp, ready to give warning and do battle with any army troops that attack."

"What was it like to be with Kony?"

"He was kind to me, promised me a pretty young wife when I was a year or two older. I looked forward to that because you can't talk to the girls in the camp, and if you do get into a relationship with a girl and it's discovered then you're both killed the same day by firing squad."

"What does he look like?"

"He's average height, not tall, not short, and he's not slim or fat. He usually wears camouflage uniform, but when he preaches he puts on a long robe like an Arab."

"What about his many wives?"

"He has about fifty wives, with more than twenty girls being trained by his wives to be married to him when they have their first period. He has a senior wife, Fatimah, and she's about twenty-nine. She told me she's been with Kony for seventeen years. All his other wives respect her, and each night when Kony chooses the wife he'll sleep with, Fatimah goes and gets her."

"What about the terrible punishments I've heard that the LRA leader metes out for relatively minor offences? Did you ever see Kony punish anyone?"

"Yes. For instance, he ordered that adultery or having sex while not married were serious sins punished with death by firing squad. We were all

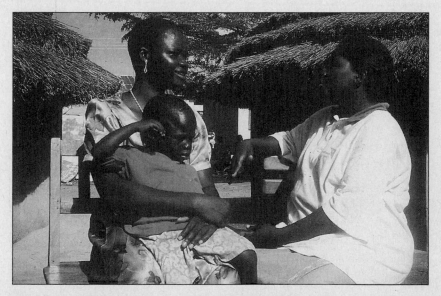

One of the child-mothers rescued from the LRA sits with her son and counselor Jacqueline Okongo at the Children of War Rehabilitation Centre. The LRA leader, Joseph Kony, had forced her to become one of the many wives of one of his commanders.

made to watch each time it happened, but Kony stayed in his house during those times. He never told me why."

"Did he ever go with you on ambushes of army troops?"

"No, never."

"When you were abducted, did the LRA soldiers make you kill and eat any of your friends on the march through the jungle to their camp?"

Peter's eyes cloud with sadness, as tears swiftly well in his eyes. He turns away, as if embarrassed by this show of frailty of spirit. "Yes," he says softly. "I was made to do those terrible things, and I'm now trying to forget them. But at night in my sleep I'm back in the jungle camp with Kony, and I often wake up crying for help."

We depart for Kampala the next morning soon after dawn because we need to reach the capital by noon. My plane leaves at mid-afternoon for Johannesburg to connect with a flight to Sydney. Jane is unusually quiet as we pass by the turnoff to the children of war center, and then head toward Kampala on an almost empty road. "At this time of day, it's still dangerous to travel by road until we reach the Nile because the LRA have infiltrated this area and they could ambush us," she says so softly that I barely hear her.

When we reach the Nile about an hour later she lets out a sigh of relief, knowing that once we cross the bridge we are out of LRA territory and out of danger. "Those poor children, those poor children," she murmurs. "Why has their suffering been ignored? The UN sends troops all over the world to fight evil, in Darfur, Kosovo, the Congo, and yet turns its face away from the children of Acholi. Why?"

I know it is not a question for me, because she murmurs the words to herself while staring through an open window at the passing jungle. But if it were meant for me I could give her no answer.

La Muerte es Grande

CANNIBALISM AND THE BARBARIC AESTHETES

CHAPTER 20

"They [the Aztec priests] strike open the wretched Indian's chest with
flint knives and hastily tear out the palpitating heart which, with the
blood, they present to the idols in whose name they have performed
the sacrifice. Then they cut off the arms, thighs, and head, eating the
arms and thighs at their ceremonial banquets. The head they hang up
on a beam, and the body of the sacrificed man is not eaten
but given to the beasts of prey."

Bernal Diaz del Castillo, a soldier of fortune who fought under the
command of the Spanish conquistador Hernán Cortés, witnessed
that ritual—a daily one—atop the Aztecs' paramount temple in Tenochtitlán, where war captives were sacrificed to the Sun God, Huitzilopochtli.
The blood from the victims' hearts and bodies—the person's very life-
force—soaked the deity, its immense power banishing darkness from Earth
for yet another day.

Diaz and Cortés were fighting an ultimately successful campaign to
topple the mighty Aztec empire in Central America beginning in 1519,
and even Diaz, from a regime famous for its bloodthirsty Inquisition, was
shocked by the ritual. After the priests sent the body tumbling down the
steep steps, the head was severed, stripped of skin and flesh and placed in
a skull rack by the temple in a grisly display of Aztec power over their enemies. The limbs were hacked off by butchers and given to the warrior who
captured the victim to eat in a ceremonial stew with maize, called *posoli*, at
a banquet with his family and friends.

This ritual cannibalism was a common sight for the Spanish invaders.
Cortés, in a lengthy letter to his sovereign, Emperor Charles V of the Holy
Roman Empire, wrote of one encounter: "It happened however that a Span-
iard found an Indian of his company, a native of Mexico, eating a piece of

flesh of the body of an Indian he had killed when entering that town, and this Spaniard came to tell me about it."

The Spanish were ruthless warriors, well practiced in atrocities to shatter the fighting spirit of their foe, but even they blanched at the horrific human sacrifices they witnessed as part of the Aztecs' everyday life, and the cannibalism that sometimes followed. This was no fanciful fiction, dreamed up by the Spanish invaders to justify conquest and the harvesting of slaves. The greatest chronicler of Aztec life, Bernardino de Sahagun, a Spanish friar who diligently learned the Aztec language, Nahuatl, to better seek the truth about Aztec life, was given numerous descriptions by Aztec nobles of their people's ritual cannibalism.

☠

These gruesome descriptions of man-eating come from contemporary accounts I have been reading on the long plane journey from Sydney to Mexico City, including Diaz's journals and Cortés's letters home to his ruler. After the horror of mass cannibalism in modern-day northern Uganda, I am now on my way to a land on the other side of the globe that suffered from extreme cannibalism in the past.

Mexico City is the home of twenty million people, as many people in one valley as in my home country which is the size of the continental United States. A taxi takes me from the airport to Zocalo, the capital's historic center, through streets bustling with traffic and pedestrians. The traffic policemen look as if they are on duty in the dangerous streets of Baghdad, armed with pistols and sometimes assault rifles, and with flak jackets strapped to their chests. The driver, Miguel Velasquez, shrugs when I ask why traffic cops are so heavily armed. "Mexico City is a dangerous place, and they need all the protection they can get. Our gangsters are well armed, and there are shootouts every day."

A newspaper is on the seat next to him, and he passes it to me. The front page is filled with a bold headline and just one large picture: a gangster sprawled on the pavement, his body riddled with bullets. His face has almost been demolished, just a fleshy pulp with no eyes or mouth, and he looks as if his clothes have been soaked in red dye. A pool of blood has formed around him.

"What was his crime?"

Miguel shrugs again. "Who cares. He lived dangerously and died young. He probably didn't get a chance to go for his gun. The cops here shoot first and ask questions afterwards, but by then the gangster is usually dead. And, the police also wear bulletproof jackets to protect themselves from other cops. They're always having fights. It's usually some crooked drug deal they were arguing over, but sometimes a bunch of off-duty cops go to a bar, they drink too much tequila, they start arguing and—Bang! Bang!—they start shooting each other."

The capital of Mexico is a city soaked in blood in past centuries and, it seems, now. In the center of the colonial precinct where I am headed are the remains of the temple where Diaz saw the Aztecs sacrifice humans to their sun god. We pass a park, the Alameda, Mexico's first, built in 1592. In earlier times it was an Aztec market. The fountains, statues and pathways are illuminated by colored lights, but there are enough shadowy nooks and crannies for me to see that any night-time visit could be risky. Alameda was in the colonial *ciudad,* or city, and in the sixteenth century the priests in Mexico who ran the grisly witchhunt for heretics known as the Inquisition used Alameda for public executions. Watched by large gatherings, the priests tied their victims to a stake and burned them.

The breeze swishes by my taxi's open window: car engines growl, tires squeal and horns honk. With enough imagination, you can hear the cries of the Aztec vendors selling tortillas and fish mingled with the screams of the priests' victims as the flames licked at their feet.

When Cortés was here the Aztec capital city was built on a small island on Lake Texcoco and could only be reached by narrow causeways connecting it to the mainland. He wrote to Emperor Charles: "In this plain there are two lakes which cover almost all of it, for a canoe may travel fifty leagues around the edges. One of these lakes is fresh water, and the other, which is the larger, is of salt water. This great city is built on the salt lake and no matter by what road you travel there are two leagues from the main body of the city to the mainland. There are four artificial causeways leading to it, and each is as wide as two cavalry lances."

The city's isolation would nearly prove to be fatal to Cortés. When Montezuma II, the emperor of the Aztecs at the beginning of the Spanish conquest, died in Cortés's custody the enraged Aztecs attacked and trapped the Spanish. The conquistadors fought their way out of Tenochtitlán at night along one of the causeways with great loss of life, though Cortés survived. Safely on the mainland, he wept beneath a tree at the defeat, and the rout became known as *La Noche Triste*, the Sad Night.

Diaz noted that in one temple at Tenochtitlán the Aztecs had a zoo, keeping "all kinds of beasts of prey, tigers, and two sorts of lions and beasts rather like wolves, and foxes and other small animals, all of them carnivores." The Aztecs also kept deadly snakes there. "We know for certain, too, that when they drove us out of Mexico [Tenochtitlán] and killed over eight hundred and fifty of our soldiers, they fed those beasts and snakes on their bodies for many days."

In his monumental book *Conquest: Cortes, Montezuma and the Fall of Old Mexico,* published in 1995, Hugh Thomas tells of the sacrificed captives' fate:

> One must assume that afterwards, as was always the case, the hearts of these men, Spanish and Indian alike, were placed in the stone 'eagle bowl' and that the captor dined off one of the thighs, while the other was eaten in the palace. If there were several captors, as there may have been, 'on the bridges,' the bodies of the captives taken were divided. The first of the captors took the right thigh. The second captor took the left. The third took the upper right arm. The fourth took the left one. The fifth took the right forearm. The sixth took the left forearm. Probably the torsos were either handed over to the animals in the zoo or taken for consumption by vultures on a remote part of the lake. Their heads, and the captured horses' heads too, would, of course, be displayed in the skull rack."

Cortés, in revenge, in July 1521 led his troops and thousands of Indian allies back to sack Tenochtitlán. An Aztec elegy for the fallen city read, "Broken spears lie in the roads; we have torn our hair in grief. The houses are roofless now, and their walls are red with blood."

The lakes have long since been artificially drained, and the reclaimed land used to house most of the city's bulging population. The narrow streets of the historic center are lined with balconied stone buildings of three or four stories that would not look out of place in eighteenth century Madrid. Dominating the old city is the enormous Plaza de la Constitucion, the world's second-largest public square. It is just after nine in the evening, and the square is thronged with people watching an outdoor concert. Across the road looms the shadowy cathedral. The Templo Mayor, the great temple of the Aztecs, or what is left of it, is nearby. It had been a long plane journey and, despite the seductive throb of Indian drums in the plaza and the nearness of the great temple's ruins, I check into a hotel across the road from the cathedral and soon fall asleep.

☠

At ten the next morning the huge paved plaza once again swarms with visitors, mostly Mexican tourists. The locals' fascination with their ancestors is still strong and about three hundred women, children, and men gape at five Indians wearing loincloths and Aztec-style headdresses made from peacock plumes. These are poor imitations. In contrast, the vivid tropical colors of the Aztecs' native birds made for spectacular headdresses. The most sought after were the iridescent green feathers of the quetzal, and these made up most of Emperor Montezuma's imposing surviving headdress, though the blue feathers of the cotinga and numerous small gold disks were also used. It took the feathers of about 250 birds to make Montezuma's crowning glory.

To my inexpert eye the dancers' simplistic performance, dancing on the spot in a circle to drums, seems a fake, unworthy of the Aztecs. The dancers are often out of step and sometimes stare at the leader, a middle-aged man, looking for a hint as to what they should do next. The tedium is occasionally relieved by a signal from the leader that prompts the dancers to shuffle toward each other for a few moments before reforming the circle.

Nearby, an Indian shaman in his fifties and clad only in a loincloth and moccasins stands by a smoking brazier filled with burning incense. He has high cheekbones, lithe muscled body, tea-colored skin, narrow eyes and a thick shock of salt and pepper hair. For a dollar the shaman blesses me in what I trust is an ancient ceremony, swathing me with the smoke by waving the brazier around my head, between my spread legs, about my body and along my limbs while chanting what he tells me is a healing song.

Along the pavement in front of the cathedral, vendors with broad swarthy Indian features sell postcards, prints and small plaster statues of Jesus and Mary. The Catedral Metropolitana's construction began in 1573, and it is the largest church in Latin America, though its huge size could be its death warrant, as it is slowly sinking into the soft marshy earth beneath it.

Inside, scaffolding acts as a skeleton to prevent the stony structure from collapsing. Behind the altar stands a tall golden backdrop, and huge paintings of Jesus, Mary and the saints hang high on the walls. St Joseph is a notable omission as he often is in churches. I wonder what place he holds in heaven, before the throne of the Almighty with his wife, Mary, and Jesus his borrowed son, or tucked away, perhaps happier, in a carpenter's den making beds and dining tables for heaven's new arrivals.

On a narrow cobblestone street at the back of the cathedral, within sight of the ruins of the Aztecs' great temple, I pass several shops selling life-size statues of Jesus and Mary. An organ grinder in khaki uniform and cap winds out a mournful tune as I approach the sacred ceremonial ground of Tenochtitlán, with the 114-step-high Templo Mayor, the paramount temple, once at its center. But now, very little of the temple and the surrounding sacred edifices can be seen. All that remains is a scattering of ceremonial mounds, relics of one of ancient Mesoamerica's most impressive constructions. The Spaniards tore most of it down because it was to them a demonic abomination, and because they wanted the stones to build the first cathedral next door.

On an earlier visit to Mexico City I had seen a reconstruction of the temple complex set in a large light-filled room at the National Museum of Anthropology. The great temple soared high above the sacred precinct with its steep steps designed so that war captives and slaves sacrificed at the top would tumble to the bottom without snagging on the steps once the priests hacked out their living hearts. The steps led to the twin shrines of Tlaloc, god of rain, and Huitzilopochtli the Aztec sun god and also the god of war, who sprang from his mother's belly fully grown and armed with a knife made of obsidian. The war god's shrine was topped with racks bearing thousands of skulls of sacrificial victims, a graphic display of Aztec might, and a salutary warning to any who dared defy them.

The Aztec gods' thirst for human blood and hearts was insatiable. In 1487, when the great temple was rededicated after it was rebuilt, more than twenty thousand war captives and slaves were sacrificed there to the gods over a period of several days. From the north, south, east and west of the capital, warriors herded enormous queues of victims shuffling along the causeways and the streets to their grisly deaths. Archaeologist Richard E. W. Adams of the University of Texas in San Antonio wrote of the era that "the Aztec . . . had also become a mad world of bloody terrorism based on the cynical, psychopathic policies of the high imperial rulers. Coronation ceremonies of the later kings were accompanied by the offering of fantastic quantities of human victims to the gods."

Atop the pyramid the Aztec emperor, Ahuizotl, murdered the first victim. He sliced open the chest with a sacred obsidian knife bearing an eagle warrior's image as its handle, and tore out the still beating heart to be ritually burnt as an offering to the sun god. The corpse was then sent tumbling below to the butchers. When the Emperor and other Aztec leaders tired of doing the killing, scores of priests took over the ghastly task.

Ahuizotl was the Aztecs' most victorious warlord, a ruthless conqueror much loved by his warriors in the same way that Julius Caesar won the hearts of his legionnaires by being very visible on the battlefield. Ahuizotl planned this horrific slaughter to impress forever more on tributary states and enemies the empire's enormous might. As most victims were war captives, following ritual practice their captors would have feasted on their severed limbs in what might have been history's greatest orgy of cannibalism.

The great temple was flanked on both sides by minor temples, and beyond by schools for warriors and palaces, including the Emperor's mansion. The reconstruction at the museum portrayed the white stucco buildings as pristine. It was nothing like that according to Alejandro Terrazas Mata, a Mexican physical anthropologist at the National Autonomous University of Mexico. "The complex's most sacred area resembled a slaughterhouse because the temples where humans were sacrificed every day were coated with the victims' blood, inside and outside." It would have been sacrilegious to wash it away, he says, and for that reason it was also left to cake the clothes of the priests who slaughtered the victims.

Cortés referred to this holy gore in a detailed letter to his sovereign back in Castile. "I had those chapels where they [images of the Aztec gods] were cleaned for they were full of the blood of sacrifices." There were about five thousand priests in Tenochtitlán, nightmarish figures, just like their gods. Clad in black bloodied robes they "never comb their hair from the time they enter the priesthood until they leave." Their hair, never washed, was stuck together with dried sacrificial human blood. They pierced their tongues, legs and even their penises with sharp thorns and rods to draw sacrificial blood. Ever balancing the horrific and the aesthetic, the Aztecs compared the priests to hummingbirds, and the blood of sacrificial victims to the nectar taken from a flower.

Diaz was sickened by the charnel-house look of the most sacred temples at Tenochtitlán. "There were some smoking braziers of their incense, which they call *copal*, in which they were burning the hearts of three Indians whom they had sacrificed that day; and all the walls of that shrine were so splashed and caked with blood that they and the floor were black. Indeed the whole place stank abominably. . . . This Tezcatlipoca, the god of hell, had charge of the Mexicans souls, and his body was surrounded by figures of little devils with snakes' tails. The walls of this shrine were also so caked with blood and the floor so bathed in it that the stench was worse than that of any slaughterhouse in Spain. They had offered that idol five hearts from the day's sacrifices."

On the conquistadors' march on Tenochtitlán, at a town called Cholula, Diaz encountered a horrific sight. "I cannot omit to mention the cages of stout wooden bars that we found in the city, full of men and boys who were being fattened for the sacrifice at which their flesh would be eaten. We destroyed these cages, and Cortés ordered the prisoners who were confined in them to return to their native districts. Then, he ordered the [leaders] of the city to imprison no more Indians in that way and to eat no more human flesh."

For a long time critics cast doubt on such reports by Diaz, claiming they were fanciful inventions intended to demonize the Aztecs. But Diaz was vindicated in 2006 when a Mexican archaeologist, Enrique Martinez, released his findings at a place about sixty miles from Mexico City called Tecuaque, or the place "where people were eaten." Cortés gave it this name when he heard about the fate there of some of his soldiers and their camp followers.

Martinez sold the rights to a British TV company and has since fallen silent about his discoveries, but before the sale he showed a Reuters reporter, Catherine Bremer, around the site. Bremer wrote that "skeletons found at an unearthed site in Mexico show that Aztecs captured, ritually sacrificed, and partially ate several hundred people travelling with invading forces in 1520. Skulls and bones from the site show about 550 victims had their hearts ripped out by Aztec priests in ritual offerings, and were dismembered or had their bones boiled and scraped clean."

The Aztecs captured and murdered a slow-moving caravan made up mostly of porters and cooks who were Maya Indians and mestizos, or mixed-race people, and the Spanish soldiers accompanying them. Bremer learned from Martinez that "the prisoners were kept in cages for months while Aztec priests selected a few each day at dawn, held them down on a sacrificial slab, cut out their hearts and offered them to Aztec gods." Martinez told her that "it was a continuous sacrifice over six months. While the prisoners were listening to their companions being sacrificed, the next ones were being selected."

Martinez explained that, following Aztec practice, the priests and town elders sometimes ate their victims' raw bloody hearts or cooked flesh from their arms and legs once it dropped off the boiling bones. He has found knife cuts and even teeth marks on the bones that showed which ones had the meat stripped off to be eaten.

Alejandro, too, has proved that cannibalism was commonplace among the Aztecs, using forensic methods to study human bones he dug up from their ancient garbage dumps. "I've found human bones that were clearly used as tools, and other human bones that we found in the kitchens and which bore characteristic cut marks on arm and leg bones. They came from butchered bodies and had been boiled in the cook pots."

He is now researching cannibalism among the inhabitants of Mesoamerican states including Teotihuacan, the ancient city near Mexico City whose way of life, about fifteen hundred years ago, was largely carried on by their cultural ancestors, the Aztecs. That bequest was one of the bloodthirstiest in history. The rulers and their willing accomplices, the priestly caste, of many Mesoamerican states for almost two millennia cowed their enemies and their own people with gory reigns of terror that wove together continual warfare, human sacrifice in vast numbers never seen before or since in history and ritual cannibalism.

The American anthropologists Christy G. and Jacqueline A. Turner, in their book, *Man Corn*, wrote: "When one looks at the entire picture of Mexican warfare, religion, human sacrifice, public and private trophy taking, and cannibalism, one senses that a very powerful, dehumanising sociopolitical and ideological complex had evolved in central Mexico, even before the time of the monumental constructions at Teotihuacan, at least 2,500 years ago. Time and again this complex was overthrown or collapsed, only to rise again and grow ever more powerful."

Some experts, however, continue to assert that the Teotihuacans were not cannibals, that the man-eating emerged among much later Mesoamerican societies such as the Aztecs, and Alejandro has promised to show me why they are wrong. After a visit to Teotihuacan to see the ancient city, I am to visit him at his lab.

My day at the Templo Mayor over, I head for dinner. Afterward I stroll across to the plaza under a darkening sky to watch a concert by a buxom popular singer. About five thousand spectators stand shoulder to shoulder, entranced by her bravura performance. At the end the singer raises a plump fist and roars into the microphone, "*La muerte es grande!*" or, "Death is great!" The crowd shouts back its agreement. It seems an eerie sentiment to exalt, especially at a pop concert, but perhaps it was prompted by the bloody history of this city, from Aztecs to conquistadors to gangsters and cops. You must have to glorify *la muerte*, or at least come to a willing accommodation with it, when so much of your past and present has death as a central theme.

CHAPTER 21

I take breakfast on the hotel rooftop which overlooks the Plaza de la Constitucion. The biggest flag I have ever seen flutters from a flagpole in the square, a bold statement of Mexican nationalism. It strikes me as apt because we Australians were given independence by our British overlords with hardly a drop of blood spilled, while the Mexicans had to shed so much to win their independence from Spain. Perhaps this is why we do not place our hands over our hearts when the national anthem is played, and why at an international sporting contest, you are more likely to see Australians waving flags emblazoned with the boxing kangaroo—a sturdy fellow with a cocky glint in his eye and with the gloves raised and ready for the fight—than you are to see the national flag, a star-spangled Southern Cross featuring as an insert in a corner Britain's Union Jack.

It is 9 am, and the plaza swarms with visitors. The pretend Aztec ceremonial dancers are already here, spinning in a mindless circle as one of them wanders through the gathering audience collecting tips. At the cathedral, visitors are streaming into its cavernous interior, for Mass or just to gape.

At mid-morning I leave for the ancient ruins of Teotihuacan, thirty miles to the northeast. The taxi passes by the cathedral, where about one hundred men are lined up along one of its walls. They hold hand-written signs against their chests and stare grimly into space. "They're unemployed and looking for a day's work, and the signs state what they do," says my guide for the day, Carlos Gonzalez, a short squat man with the swarthy features and thick jet black hair that hint at Indian ancestry. "There are laborers, electricians and others."

With their backs facing the same ancient wall and lined up in ranks are about two hundred paramilitary troops clad in full riot control gear. Their legs, arms and torsos are swathed in bulletproof armour while their heads are shielded by helmets with transparent visors. They carry batons and pistols. There seems not one spark of humanity in any of them, their stares are blank, not a hint of emotion, and they look like robot law enforcers from a violent Japanese *manga* cartoon. Carlos dismisses them with a contemptuous glance. "They're the *Policia Federal*. A head of state is visiting the square today, and they're here to protect him, but from whom I don't know."

The roads are gridlocked, too many cars trying to squeeze through too many narrow streets built in the horse and buggy days. As we crawl along I am surprised to see that Mexico City is the cleanest city I have encountered in my journeys to more than one hundred countries. The impression does not change throughout my week-long stay. I cannot remember spotting a single piece of garbage on the streets, which swarm with an army of cleaners carrying pails and buckets at all times during the day and into the night.

It takes forty minutes to clear the suburbs, and then we are in the green countryside. The volcanic soil bursts with fertility, and maize fields line both sides of the road. For millennia it has been the staple food of Mesoamericans, as it remains today. Ahead, backed by a large mountain, are two high stone pyramids. "Teotihuacan," Carlos announces.

The pyramids, like those in Egypt, are immense, mysterious, shimmering with history, beckoning, and I am eager to reach them, but Carlos orders the taxi driver to make a detour and we turn into a parking lot by the side of a shop. "It's the best place for souvenirs in Teotihuacan," he says.

"Carlos, I'm not interested in wasting time on souvenirs. Let's go to the pyramids."

His eyes plead with me. "All visitors come here," he says. I understand his concern: he is probably paid only a percentage of his fee by the agency, and must try to make a decent wage by taking a cut, with the owner's connivance, of whatever visitors spend at the shop. Inside the shop is a display of obsidian carvings of Aztec nobles and warriors. The workmanship is expert, but the carvings lack any sense of love by the artist for his handiwork. "Obsidian is volcano glass, and was used to make weapons by the early Mesoamericans including the Teotihuacans and the Aztecs," Carlos explains. "They used it for spearheads, arrowheads, axes, knives and war clubs. When it's sharpened it can cut off a man's head."

He picks up a dagger, its deep black surface glowing with inner light. "This is a copy of a knife used in human sacrifice," he says. "The Teotihuacans rose to become a rich and powerful nation because they had control of the obsidian mines in the mountains. They traded it as far away as Guatemala and Durango."

I take pity on Carlos and buy an obsidian rabbit for my daughter. Her Chinese year sign is the rabbit, and for more than thirty years, wherever I have gone in the world, I always try to buy her a local toy rabbit. So, the bookshelves and cupboards of her bedroom and study teem with a multicultural array of rabbits, from Rwanda to Rio de Janeiro to China itself.

At the site a broad avenue runs for just over two miles, from the pyramids to the *ciudad*, or city, and is lined with the ruins of dozens of small temples and platforms. The avenue is named Calzada de los Muertos: Avenue of the Dead. "We know from archaeological discoveries that the people who lived here believed that when they died their spirit moved down the avenue on its way to the afterworld," Carlos tells me.

Carlos takes me the opposite way, from the edge of death to the fullness of life. The grassy avenue leads to the pyramids, the enormous temple of the Sun and the much smaller temple of the Moon, with scores of steep steps leading to each summit. Teotihuacan is 7482 feet above sea level, and the oxygen here is thin. I am not yet acclimatized to the height and must breathe deeply as I walk along the avenue.

The temple of the Sun is closest, and hundreds of people are climbing to the summit. They look like ants scaling a sand castle. The pyramids soar over all the other Teotihuacan buildings. "The temple of the Sun is 210 feet tall, and is the world's third largest pyramid," Carlos says. "It's about half the size of the great temple in Giza, though the perimeter of its base is almost the same size.

"Very little is known about the city and its people, and even the name they gave it is unknown," he adds. "Teotihuacan is Aztec for 'place of the gods,' and when they discovered it many centuries after its fall they were so awed by the size of the pyramids and the complex that they believed this was where the gods were born, and then gave birth to our world. Before Teotihuacan there was only silence and darkness in the world."

Bernardino de Sahagun, that assiduous chronicler of Aztec beliefs, learned from his Aztec informants their version of the creation story. "It is told that when all was in darkness. When yet no sun had shone and no dawn had broken, it is said—the gods gathered themselves together and took counsel among themselves here at Teotihuacan. They spoke: they said among themselves: 'come hither oh gods! Who will carry the burden? Who will take it upon himself to be the sun, to bring the dawn?' The gods set fire to an enormous bonfire. Two of them then threw themselves into a bonfire, and were reborn as the sun and moon, and our World took shape."

The city was set on a grid with the ghostly avenue on an east-of-north to west-of-south axis, and the streets are broad and narrow, perfectly straight and intersecting the avenue at right angles. This geometric master plan provided the model centuries later for Aztec cities, including Tenochtit-lán. First in our path is the Temple of Quetzalcoatl, the feathered serpent god, also a favorite deity of the Aztecs. He is god of the wind, and also art and knowledge. The temple is set in a large grassy square, and the base is guarded by fierce toothy serpent heads that rear out of the stone, still capable of prompting shivers of fear among visitors more than fifteen hundred years after they were carved.

Alejandro, the anthropologist, had told me that although much smaller than the two pyramids, the temple of the feathered serpent was the most powerful in the complex. Many historians claim that the Aztecs believed Cortés was the reincarnation of Quetzalcoatl, who was traditionally por-trayed as having fair skin, a beard and coming from the east, just like Cortés. But once they witnessed the brutal way he crushed their armies, they realized that no Aztec god would so harm his people. Nearby are the Palace of the Jaguars, the Butterfly Palace, and the Temple of the Feath-ered Conches.

All the buildings come from what archaeologists call the classic period of Mesoamerican history. Teotihuacan is so significant that it defines the span of the classic period, from about 100 BC until the city's fall around AD 750. "At its height in the sixth century, the city covered fifty square kilometers and was one of the world's largest cities, with a population of about two hundred thousand," says Carlos. "It was bigger and more tech-nologically advanced than any European city at the time."

The temples and pyramids have long been stripped bare of their decora-tions and murals. All that remains is the underlying stone, the skeletons of one of the most important and yet mysterious cities in history. At the apex of its power the exteriors of the many sacred buildings were lined with pink stucco so that the entire complex shone in the sunshine like a precious jewel. "Inside the temples, palaces and houses, the walls were painted red because that's the color of blood, our life-source and the life source of the Aztec gods," Carlos says as we near the Temple of the Sun.

In the middle of the eighth century, something terrible happened here, and someone—whether invaders or rebel Teotihuacans—set fire to all the temples and tore them down. This destruction and probable massacre gives the complex an eerie feeling, as if it were a city of ghosts. After the fall, the city withered and died.

The very best of the relics recovered from Teotihuacan are on display at the National Museum of Anthropology back in Mexico City, but at the base of the Temple of the Sun is a small museum whose most interesting exhibit is a recreation of the burials of sacrificed humans. Several skeletons are on display, which archaeologists believe were victims buried inside the pyramid as a sacrifice to the gods during its dedication ceremony, which took place in the second century AD. They lay as archaeologists found them inside the pyramid, side by side, sets of bones that were once humans, with necklaces strung with miniature imitation human jawbones about their necks.

Carlos, unlike Alejandro, doesn't believe that the Teotihuacans were cannibals, and so I am eager to visit the anthropologist to see what evidence he has to rebut my guide's claim.

☠

I had vowed to avoid churches, not because I am irreligious but because I preferred to focus my attention on the Teotihuacans and the Aztecs. But this is Mexico, and one particular Catholic church was crucial to the Spanish conquest. On the way back to Mexico City, Carlos takes me to Latin America's holiest shrine, the Basilica of the Virgin of Guadalupe. Mexico's patron saint, she is the object of intense veneration. On her feast day, December 12, Carlos says millions of Mexicans throng the church and its grounds to view her holy cloak and pay homage to *La Reina de Mexico*, the nation's "queen."

In an outer suburb, he leads me into a paved square dominated by an enormous statue of Pope John Paul II, standing between a basilica begun in 1536 and a 1976 monstrosity, the Basilica Nueva. The older church,

built in Baroque style, is pleasing to the eye, but the newer church is a typically ugly example of neo-architecture. Its wood, plastic and steel makings
look as though they were thrown together in a paint-by-numbers effort, all
sharp angles and spiky perspectives and blockhouse weightiness without a
hint of grace, balance or beauty.

The two shrines were built to commemorate a signal event in Mexican
history that might be a mythical concoction transformed into truth by the
turning of the centuries and the need of the local people for just such a
miraculous happening. The story says that just before Christmas in 1531,
a decade after the Spanish overthrew the Aztec empire, an Aztec, Juan
Diego Cuauhtlatoatzin, came to the bishop, Fray Juan de Zumarraga, to
report a vision: the Virgin Mary had just appeared to him on a nearby hill.
She asked that a church honoring her be built on the hill. "The bishop didn't
believe him," Carlos tells me as we enter the new basilica, "and asked for a
miracle from the Holy Mother as proof. The next day she appeared again
to Juan Diego and told him to gather roses from the hill, even though they
were not in season. When Juan Diego found roses there and took them to
the bishop, as they tumbled from his cloak the Holy Mother's image on it
was revealed. This convinced the bishop, and we Mexicans have revered
the cloak as our holiest icon ever since. Pope John Paul came to Mexico
just to view the cloak."

Juan Diego likely did not exist, and there is no written record of the ap-
parition until 1648, more than a century after Mary supposedly appeared
before him. Even the Basilica's abbot said in 1996 that he thought Juan
Diego was a myth. I wonder whether he, the abbot, had been courageous
enough to make that claim in person to John Paul, an ardent devotee of
Mary, during one of his visits.

It is mid-afternoon on a weekday, but the church, which holds 10,000
people, is almost full. The worshippers have come from across the nation.
The Mexicans are a passionate people, and much of their abundant out-
pourings of emotion spring from their devotion to Catholicism. Life is
tough for most Mexicans, and their fidelity to religion is inspired by prayers
for a better life.

The congregation sings a hymn, led by a priest with a microphone, as
scores of devotees approach the altar slowly on their knees, eyes fixed on
the cloak draped within a gold and silver frame at the back of the altar. I
move forward for a closer look and am disappointed. It is the least believ-

able major religious relic I have ever seen, and looks like the work of a hack painter. The Virgin is easy on the eye, but without a hint of her charismatic divinity.

Looking at this "miracle" I wonder why Mary did not call on Michelangelo to paint the cloak for her, but then realize that he was more likely to be a confidant of Satan in hell. But surely there was at least one great painter, despite the many fleshy temptations put in the way of such men, who made it to heaven. The mythology seems to have the bishop recognizing Juan Diego's con job. He was suspicious of the claim, he asks for proof, and Juan Diego turns up the next day with a cloak depicting the Virgin and claiming it came from her. The Aztecs had just seen their own religion destroyed by the invading Spaniards, and perhaps the message of the story is that the bishop used the supposed vision of Mother Mary to woo the Aztecs into becoming Catholics by showing she cared enough about them to organize a miracle involving one of their own.

So many religions have elevated human sacrifice, whether real or as a metaphor, to a central place in their theology. Jesus died on the cross for his followers, and the Aztec victims died atop the pyramid for their gods. Christians even have a form of metaphorical cannibalism in Communion, with believers eating and drinking what is claimed to be the body and blood of Jesus transformed into bread and wine. The Aztec version is considerably more real and more terrifying.

Back in the hotel that night, about one hundred paces from where scores of thousands of humans were sacrificed at the great temple of Tenochtitlán, I read an eyewitness account of the sacrifices and cannibalism. An Aztec noble gave it to the chronicler Sahagun. It detailed how there was a stately religious feel to the ceremonies leading up to the sacrifices. When dawn came, the captors came for those they had captured and delivered them to the priests who grabbed the hair on their heads and led them up the steep steps to the top of the great temple.

"And when some captive faltered, fainted or went throwing himself upon the ground, they dragged him. And when one showed himself strong, not acting like a woman, he went with a man's fortitude; he bore himself like a man; he went speaking in manly fashion; he went exerting himself; he went strong of heart and shouting, not without courage nor stumbling, but honoring and praising his city. He went with a firm heart, speaking as he went: 'Already here I come! You will speak of me there in my home land!'"

At the top of the temple the captive was spread-eagled on the back over the sacrificial stone by six priests including the executioner. He swiftly cut open the breast with a wide-bladed flint knife and dragged out the still beating heart, called the "precious eagle-cactus fruit." He held it up to the sun to nourish the god and then placed it in a vessel.

"Afterwards they rolled (the captives) over; they bounced them down; they came tumbling down head over heels, and end over end, rolling over and over; thus they reached the terrace at the base of the pyramid."

The butchered bodies were carried to the homes of the captors to cut up and divide the various parts. The Emperor, Montezuma, was given a thigh from each body. The captor cooked for each guest at the feast a bowl of stew of dried maize which also contained a piece of the captive's flesh.

CHAPTER 22

The morning begins in the historic precinct at the imposing Palacio de Belles Artes, built in 1901, whose massive slabs of white marble resmble the kind of overblown wedding cake favored by Greek immigrants in my country. Inside is a riot of Art Deco paintings, statues and friezes. I peek through a door into the large theatre to see the art center's masterpiece, the Tiffany glass stage curtain, a striking depiction in translucent panels of the volcanoes that loom over the city and the valley.

I have come to view a major exhibition of the work of painter Frida Kahlo, whose tortured life was made into a Hollywood movie starring the Mexican-born actress Salma Hayek not long ago. The exhibition is being staged to celebrate the hundredth anniversary of Kahlo's birth, and the line for tickets extends down the ornate staircase and out into the street. Kahlo is Mexico's most famous artist internationally, but as I wander among the paintings, none of them great and many quite ordinary, it seems to me she is honored more for her made-for-movies life story than the genius of her art.

Kahlo suffered terrible physical and mental pain throughout much of her life, and this slipped into a psychopathic obsession with her own image and the repeated depiction of the pain she had to bear stoically. Half her paintings were self-portraits. At the exhibition, painting after painting has Kahlo staring out at the viewer, her grim face distinguished by thick eyebrows joining in the middle and by a moustache. Kahlo once explained, "I

249

paint myself because I am so often alone, and because I am the subject I know best."

Her features in the portraits are always twisted into a scowl, as if she is forever passing judgment on fate for having condemned her so cruelly. At age six, she contracted polio, which left her with one leg shorter than the other. But the worst physical pain came when she was in a bus hit by a tram when she was nineteen, and the impact broke bones all over her body. A steel handrail thrust through her pelvis. She suffered terribly for the remainder of her life, and was often bedridden.

The intense mental pain came from a philandering husband, painter Diego Rivera, a world-famed muralist, notorious for his many affairs including one with her sister Cristina. Kahlo was besotted with him, even though he was obese and had a frog's face, and she divorced and then remarried him. I suspect that more than her middling paintings, it is that stoic bearing of so much agony that endears Kahlo to Mexicans. The richest man in the world may now be Mexican, Carlos Slim having overtaken Bill Gates in 2007, but most Mexicans struggle to afford a decent life, and it could be that Kahlo's life-long suffering strikes a sympathetic chord among them. That, and her fevered obsession with suffering and death. *La Muerte es grande!*

Out in the heavily polluted air once again, the dark mood of Kahlo's paintings lifts with my first glimpse of sunshine through the gasoline haze. Mexico City nestles in a basin almost surrounded by hills that block much of the wind that would otherwise blow away the pollution. Across the street in Alameda lovers are entwined in passionate embraces on park benches, a mariachi band is pouring its over-pumped heart out beneath the trees to a few listeners and laughing children eat ice cream as they perch on the rim of a splashing fountain.

☠

During Aztec times Tenochtitlán's two square miles of land was surrounded by a lake, Texcoco, but almost all the water has been artificially drained and the land reclaimed. However, at Chapultepec, located in a Mexico City suburb, one of the world's finest public parks, amid the thousand acres of hilly greenery is a waterway where the locals go boating, especially on the weekend.

Here is where the Aztecs, then a ragged bunch of scavengers scorned by other tribes, first settled when they entered the valley in the fourteenth century. They called themselves the Mexica. One of their prophecies had told them that they would find their future home at a place where they would spot an eagle sitting on a cactus holding a snake in its claws, and in 1323 they did, at the spot that would come to be Tenochtitlán. Within a century their military genius and unsurpassed ferocity and bloodlust led them to become one of the greatest empires ever seen in Mesoamerica.

So, it is fitting that the park contains the imposing National Museum of Anthropology, even if it was designed by the same architect who planned the ugly Nueva Basilica in honor of the Virgin of Guadalupe. The museum is incomparably better, and leads the viewer on an often breathtaking journey through more than three thousand years of Mesoamerican artistic achievement in spacious well-lit galleries that flow around the large courtyard in chronological order.

The Teotihuacan gallery is dominated by a reproduction of a polychromatic section of the front of the Temple of Quetzalcoatl with its fierce toothed serpents' heads and their sinuous bodies painted in red and green, the original colors. It is a startling contrast to the bare bones look of the temple at Teotihuacan. Also on exhibit are funerary masks of high artistic merit, carved from volcanic rock and covered in a mosaic of green turquoise chips.

The Temple of the Sun's immense façade was originally decorated with numerous symbols of death—no surprise there. The museum holds one, a carved disk that resembles the sun and flaunts in the middle a gruesome human skull with its tongue stuck out. There is nothing in the gallery that hints of cannibalism, and the most gruesome exhibit is a large terra-cotta figure of a priest of the cult of Xipe Totec—his name means "our flayed lord"—the god of spring. He is clad in the flayed skin of a sacrificial victim, a representation of a most gruesome ritual.

In the Aztec religious calendar, the coming of spring was marked each year by a ceremonial period running from the 6th to the 25th of March known as the Flaying of the Men. War captives selected for the ceremony were given a feather sword they had to use to defend themselves against warriors armed with obsidian swords. Once they were slain, as they inevitably were, expert skinners would remove their entire skin, from toe to head, in one piece, from each captive. The cult's priests then slipped into the skins and wore them for the entire season. The skin was thickened by dried globules of fat and blood. The Teotihuacans revered this ritual as a symbol of regeneration in springtime, the renewal of the earth's skin, and the Aztecs carried on the tradition, linking the Flayed Lord with the east from where the blood-red sun rose every day.

The priestly terracotta figure in the museum has the nightmarish look I found in all the sculptures and paintings of the ritual. The priest's mouth would have been barely visible through the stretched skin around the flayed victim's mouth and lips, and the captive's hands and feet, still attached, would have flapped about in a ghoulish way as he danced. The Flayed Lord himself was depicted, clad in an elaborate headdress and wearing the flayed skin of a captive, in the Codex Borgia, an illustrated religious manuscript and one of the most important documents to survive the Aztecs. It was painted by Aztec scribes decades before Cortés overthrew the empire as a guide to important ceremonies, as a visual catalog of supernatural beings, and as an outline of the Aztecs' 260-day ritual calendar (which was separate from their 365-day solar one). The Codex was also used for divination and ritual prophecies, and is now held by the Vatican.

The museum's main gallery is devoted to the Aztecs. Mammoth rock statues demonstrate the Aztecs' favorite themes: death and destruction. Near the entrance is a stone receptacle that held the hearts of sacrificial victims. The large carving of the sun god's mother, the goddess of nature, flaunts a skirt of serpents and is carved with symbols of human sacrifice, include a grinning skull. She wears a necklace of human hearts and severed hands. Here too is that pristine reproduction of the sacred complex at Tenochtitlán, while a superbly carved figure of the god Xochipilli, looking like a guardian of hell, has him sitting on a throne of flowers and butterflies sipping nectar. Immense beauty and utter terror in one vision, one moment.

One of the most dramatic exhibits is the enormous stone disk—weighing twenty-four metric tons and measuring twelve feet in diameter, found in 1790 beneath the Plaza de la Constitucion— known as the Aztec Sun Stone. Scholars once thought it represented the Aztec calendar, but now know it to be a mammoth sacrificial stone, on which captives were held by several priests at the great temple while the executioner tore out their hearts.

At the disk's center is the fearsome face of a man thrusting out his tongue, which is shaped like the broad-bladed obsidian knife used to cut out sacrificial victims' hearts. This could be either Tlaltecuhtli, the Earth Monster, or an ancient sun god known as Tonatiuh. Whoever it represents, the bloodthirsty figure grasps a human heart in each of his hands, just like the ceremonial executioner.

Outside the gallery there is more evidence of the Aztecs' obsession with death and sacrifice, large stone rings that were used as goals for their popular ball game. Following a long tradition in Mesoamerica that reaches back beyond the Teotihuacans, the Aztecs played a game called *tlachtli* with a solid rubber ball in walled courts twenty to thirty feet wide and up to one hundred and fifty feet long. The players, propelling and passing the ball with their hips, elbows, knees and heads had to shoot it through a ring up to ten feet high on either side of the court. Players on the losing team were sometimes beheaded as a sacrifice. When this deadly version was played, skulls were placed along the court, perhaps to remind the players of the outcome if they lost.

☠

A thirty minute taxi ride takes me to the campus of the National Autonomous University of Mexico, the oldest in the Americas, given its charter in 1551. The site's many buildings set in acres of lawns, shrubs and trees are dominated by the twelve-story-high library, its four sides entirely covered with mosaics depicting Mexico's Aztec, colonial and post-independence history.

Alejandro Terrazas Mata, the physical anthropologist, leads me into his small lab as a bushy-tailed squirrel scurries under the table. "She took up residence here some time ago, and believes this is her den, that she owns it, and I don't have the heart to chase her out," he says.

Alejandro looks like a hippie with his beard and ponytail. "I've been digging at a Teotihuacan site we call La Ventilla, near the pyramids, for a decade," he says, "and I've found irrefutable proof that the people there routinely ate human flesh. La Ventilla was where the middle class lived. They were the feather-makers, bureaucrats, jewelers, who served the nobles and priests, and from the number of human bones we found with clear signs of cannibalism, it seems they had access to the many bodies the priests sacrificed in the same way and presumably to the same gods as the Aztecs. After captives were sacrificed, it's possible the priests gave the bodies to these workers to eat. During the sacrifice the captive is regarded as a god, but once the priest tears out the heart and burns it, and pushes the body down the pyramid steps, it's just flesh, no different than animal flesh."

"Do you know why they ate human flesh?"

"We're not entirely sure, but I've found by studying the bones of La Ventilla's residents, a science called taphonomy, that they had very good health, very good nutrition, and so I don't think they ate human flesh for the protein. It might have been for the taste, because we found lots of cut marks on the clavicle and shoulder, and that's where you get the tastiest meat, just like goat meat today. It might also have been a way for them to feel important by copying the priests who ritually ate the flesh of sacrificial victims."

On a computer screen, Alejandro shows me some of the many hundreds of pictures he has taken of human bones he dug up in ancient garbage dumps at the site. "We found near their household kitchens many human bones that came from dismembered bodies in the same place as bones of animals they used to eat such as rabbit, turkey, deer, duck. These bones were different to the human bones they used as tools, mostly the skull and the arm bones, which were carefully disarticulated and not damaged." He shows me the top of a skull whose surface is covered with tiny indentations. "The Teotihuacans used human bones to polish leather, and also the stucco surfaces of their temples, and when you want to make tools from the bones you don't break the bones."

He shows me more bones whose color and surface is quite different. "See, these bones have been cooked: they have a characteristic waxy surface, and we found them at the kitchens of the houses. They also have different cut marks to those bones used as tools. Many have been broken, perhaps for the marrow. You don't worry about the preservation of the bone if you're going to eat the flesh from it. There's more evidence that they were deliberately defleshed. If the flesh remains on a bone, decomposition damages it, eats into it, and none of the cooked bones show that."

The next shot shows a human mandible with a few nicks found in a dump by a household kitchen. "I've seen this many times. The cut marks here show that the tongue was cut away, presumably for food."

"How can you be sure?"

"I dissected cadavers in the university lab, using an obsidian knife, the same as the Aztecs, and in defleshing the bones I had to make the same cut

marks as I've seen on the bones at the garbage dumps. Although anthropologists such as William Arens have cast doubt on Sahagun's descriptions
of cannibalism given to him by Aztec informants, the marks I had to make
on the cadavers to cut away the flesh correspond with his descriptions of
how the bodies were butchered. To convince him, I'd welcome Mr Arens to
come to my lab to see this forensic proof that the Aztecs and Teotihuacans
were cannibals."

I have brought my laptop along and show Alejandro pictures I had taken
of the Korowai cannibal brothers, Bailom and Kili-Kili, holding together
the skull of Bunop, the *khakhua* they killed and ate. They had smashed
open the skull to get at the brains. Alejandro studies the pictures carefully.
"It's the same impact point on the skull used by the Aztecs and Teotihuacans when they wanted to get at the brains. Trial and error showed both the
Mesoamericans and the New Guinea cannibals the most efficient way to
open the skull for the brains."

Alejandro had also been intrigued by Sahagun's descriptions of the ritual
of Xipe Totec, the Flayed Lord, and the Aztecs' habit of leaving the hands
and feet attached to the flayed skin of the captive. "I thought it would be
impossible to flay the victim, cutting away all the skin in one piece with the
hands and feet attached. But, in taking the entire skin from cadavers in one
piece, I found that by inserting the knife at specific points in the feet and
hands, it was easy to cut them away from the body and keep them attached
to the skin. The cut marks I made were similar to some of the cut marks I
found on human bones. It was important to the Aztecs and Teotihuacans
that they stayed attached as the priest, when he wore the skin of the flayed
person, became the god Xipe Totec, and the hands and feet make the im-
personation more real."

Although on my journeys around the world on this quest I have encountered many cannibals, and reports of cannibals, and read a large number
of historical descriptions, I never witnessed the butchering of a human
body for the cooking pot, or anyone eating human flesh. I do not doubt
the words of the Korowai, the Aghor sadhus and the children at Gulu in
northern Uganda. Why would they lie? They had no reason, and the details
they gave were compelling and consistent. But Alejandro's decade-long forensic analysis of the human bones proves with irrefutable thoroughness
that eating humans was widespread and accepted, even revered, among

at least two major Mesoamerican cultures. It makes sense that humans elsewhere on the planet throughout our long history, and even today, have eaten human flesh, both as a ritual act and for the flesh.

That night I eat a delicious meal of enchiladas and soup at a hole in the wall café near the Templo Mayor. There is no mention of chili con carne on the menu, a staple of "Mexican" restaurants back home. During the week here I eat in many restaurants and none offer chili con carne. I order Aztec soup, unsure what the waiter will bring. Alejandro had told me that the Aztecs relished a soup which featured the severed head of a sacrificed captive, and that Mexicans today carry on the tradition, though using a pig's head. I smile with relief when the soup arrives, a steaming red concoction of tomatoes, chili and a lump of melting cheese.

☠

Planning my investigation of Mesoamerican cannibalism, I had hoped to include a wider look at Mesoamerican culture, ranging north into the American Southwest. That would have me journeying on to Chaco Canyon in New Mexico, and then to Cowboy Wash near Mesa Verde in Colorado. There, the Turners and other scientists discovered defleshed and shattered bones from dismembered humans, dried human blood in cooking pot shards and fecal matter that they say prove that some of the Anasazi, ancestors of the Hopi Indians, were cannibals. In *Man Corn* they wrote "Cannibalism was practised for about four centuries, beginning about AD 900."

The Turners believed that the cannibalism stemmed from a deadly combination of "social control, ritual human sacrifice, and social pathology," tracing its origins to prehistoric Mexico—the same era and people that would give rise to the Aztecs' bloody culture.

With the violent break-up of the Teotihuacan and Toltec empires in the eighth and twelfth centuries, respectively, vanquished warrior-cultists who were dedicated to gods of human sacrifice and cannibalism might have migrated northward. The Turners suggested that some of these warriors and their families might have reached the Southwest of what is now the United States, and swooped down on the Anasazi, whose relatively large population had little central control at the time, to terrorize and conquer them. It could be drawing a long bow with no overwhelming proof, but it seems the only reasonable explanation so far of Anasazi cannibalism, a practice considered taboo by their neighbors.

But an academic censorship that works by a diktat of fear blocked the trip. Kristin Kuckelman is an archaeologist who has worked on site near Cowboy Wash, examining bones that come from cannibal victims, and is with the Crow Canyon Archaeological Center near Mesa Verde. When I phone and ask if I can come there to talk about her work, she agrees, but when I tell her that I am writing a book about cannibalism, she falls silent for a few moments. When she finds voice she withdraws the invitation saying she is not willing to talk about Anasazi cannibalism to me. "It's too sensitive," she says.

Patricia Lambert, a bioarcheologist and one of the principal scientists who investigated the claims of Anasazi cannibalism, also worked on the Cowboy Wash site. Her research on human bones found there supported the view that people there practiced cannibalism. She initially agreed for me to visit her at her university near Salt Lake City, but got cold feet when I told her about the topic.

She emailed me: "I did not know the nature of the project when I agreed to the interview—I thought it was for a Smithsonian article until I read your email to David [her colleague, an anthropology professor]. I am a North American bioarchaeologist and my research on cannibalism has already landed me in the hot seat on more than one occasion. More publicity in the popular press, especially under a title [then "Maneaters"] that many in my discipline will see as incendiary, is unlikely to do me any good and could make it more difficult for me to work, especially in the Southwest. I have read your work and think very highly of it, but I would also like to be able to continue to do my work."

It was a great disappointment—Alejandro admired her work, finding that it resembled his own findings at Teotihuacan—and would be more proof that the claims made by Arens in his book were unfounded. Perhaps what is more incendiary is that there are anthropologists who are afraid of speaking the truth when they fear academic retribution by their politically correct colleagues.

It was with no joy that I responded to Dr Lambert's email: "I understand your fear and won't press the matter any more."

This wasn't the first time I had encountered such resistance. Indeed, after I went to the Korowai, white anthropologists in Australia claimed that I had been fooled, one stating in blithe ignorance that cannibalism ended among the Korowai forty years earlier. None had ever been within hundreds of miles of the Korowai, but of course they were certain they knew better than the Korowai themselves about their customs.

EPILOGUE

Whatever the claims of the cannibalism denialists, as I board the plane for Sydney, I have one last task to accomplish. The little boy Wawa in the village of Yafufla in far-off Papua—his family had begged me to take him from the jungle to safety in the capital city of Jayapura, because his fellow villagers suspected him of being a *khakhua*, a devil-man. They believed that he had killed his mother and father by witchcraft. I saw him being treated as a pariah in the village, shunned by the other children and hated by the adults. He knew the danger he faced, and his deeply troubled eyes reflected a terrible fear. When he grew older there was a chance they would kill and eat him.

But now there is a new urgency. At home, I receive a call from Kornelius Kembaren, the guide who took me to the Korowai, who says he has just returned from Yafufla where he learned that Wawa's uncle, one of our porters, had recently been murdered at the village. Wawa's other uncles have placed a twenty-four-hour watch on him, believing him to be in extreme danger. Kornelius says they beg me to help rescue the little boy and take him to safety.

I fly from Sydney to Bali and on to Jayapura, but at the Indonesian government office the official refuses this time to issue me a *surat jalan*, the permission to enter the sensitive region where the Korowai tribe live. Without it I would be forbidden even to board the small plane to Yaniruma. Even worse, the police haul me in, question me and demand that I leave Jayapura and return to Bali by the first plane.

Somehow, the Indonesians have found out that, although I have a tourist visa, I intend to join in the rescue of Wawa and report on it. They now know that on my previous trip I did not enter the Korowai area as a tourist, as stated on my visa, an illegal act. I did it in good conscience because there had been an upsurge in rebel activity, making it difficult for a foreign writer to get permission to enter the remote Korowai territory. The senior police officer who questioned me claimed that the Indonesian government would be concerned that such a trip might be a subterfuge for contacting Papuan freedom fighters.

"Don't worry Paul, I'll go and get Wawa," Kornelius assures me. He agrees to bring along one of Wawa's uncles to spend a few months in Jayapura to help the little boy settle into his new life without fear.

I return to Sydney concerned about Wawa's future at Yafufla. But two weeks later I get a call from Kornelius and I can tell from his voice that his mission was a success. "Wawa is with me now in Jayapura," he says. "I went back to Yafufla and his family were very happy that he's now out of danger. His uncle is also with us."

As I write this, Wawa has been in Jayapura for a year. He has picked up the Indonesian language quickly and now attends school in the Papuan capital. Kornelius has just emailed me a photo of a smiling Wawa, clad in his school uniform and carrying a backpack for his schoolbooks. I put this photo next to the picture I took of Wawa when I met him in Yafufla, when he was clearly terrified, and they do not look like the same child. Wawa's uncle has returned to Yafufla, content to leave his welfare with Kornelius. I have arranged for the uncle to visit Jayapura once a year to be with Wawa so that he never forgets he is a Korowai.

In a few years Wawa will go to high school and then on to university if he has the desire and the talent. There was a risk in bringing him out of the jungle because he had to adjust to a vastly different way of life and learn a new language. But the alternative was far worse: life as a pariah in the village, a *khakhua* suspect, growing up in daily in terror, fearing that one day the *khakhua* killers would come for him.

When he is much older Wawa can go back to Yafufla to visit or even to live, as a teacher, a medic or even a doctor. By then the cult of the *khakhua* killers will probably have come to an end or be close to it, and so Wawa will be in no danger.

He and Boas are the Korowai of the future, not Bailom and Kili-Kili.

Wawa (on right) with Kornelius, who rescued Wawa from his fate as a khakhua, *at his new home.*

Joseph Kony may be in prison by then and his murderous Lord's Resistance Army long disbanded. That will leave the sadhus of the Aghor sect, Baba and Shankar in Benares, and perhaps a few hundred more spread out across India as the last living cannibal cult on Earth. We will never be free of the psychopathic cannibals, but it is probable that cannibal cults on the scale we have seen throughout recorded history such as the Aztecs and even the Korowai, will never develop and flourish again.

THE END

INDEX